Building Wireless Sensor
Networks Using Arduino

Leverage the powerful Arduino and XBee platforms
to monitor and control your surroundings

Matthijs Kooijman

BIRMINGHAM - MUMBAI

Building Wireless Sensor Networks Using Arduino

Copyright © 2015 Packt Publishing

First published: October 2015

Production reference: 1121015

Published by Packt Publishing Ltd.
Livery Place
35 Livery Street
Birmingham B3 2PB, UK.

ISBN 978-1-78439-558-2

www.packtpub.com

Credits

Author
Matthijs Kooijman

Reviewers
Anvirup Basu
Roberto Gallea
Vincent Gijsen
Randy Schur
Fangzhou Xia

Commissioning Editor
Nadeem Bagban

Acquisition Editor
Sonali Vernekar

Content Development Editor
Shali Deeraj

Technical Editor
Danish Shaikh

Copy Editor
Tasneem Fatehi

Project Coordinator
Kinjal Bari

Proofreader
Safis Editing

Indexer
Mariammal Chettiyar

Graphics
Abhinash Sahu

Production Coordinator
Conidon Miranda

Cover Work
Conidon Miranda

Layout Coordinator
Conidon Miranda

About the Author

Matthijs Kooijman is an independent embedded software developer who is firmly connected with the maker movement through a local fab lab and his work on the Arduino project. Since his youth, Matthijs has been interested in making things; for example, he built his first television remote control before the age of 10 (using a piece of rope to pull on the volume slider, not a solution that he would choose today).

Matthijs has a firm belief in the merits of open source software and enjoys contributing to the software that he uses — both by coding and helping out other users. His work experience is broad — ranging from Web development to Linux driver hacking, from tech support to various forms of wireless networking, but almost always related to open source software in some way.

About the Reviewers

Anvirup Basu is currently a student pursuing his B.Tech in electronics and communication engineering from the Siliguri Institute of Technology. Besides academics, he is actively involved in robotics, IoT, and mobile application development. Since the first year, he has been involved with Microsoft as a Microsoft Student Partner and organized three seminars and workshops on the various Microsoft technologies, mainly for Windows phones and the Windows app development.

Being enthusiastic about robotics and Microsoft technologies, he has developed several robots, both autonomous and manual, and a couple of manual robot controllers, some of which are the Universal Robot Controller for Windows PCs and Mark 1 pilot for Windows phones. He is also into computer vision and has worked on the detection of wild animals. Automated Elephant Tracker is one of his projects, in the journal named *International journal of Electronics and Communication Engineering &Technology,* under International Association for Engineering and Management Education, which includes his works on robotics and computer vision.

His website, `http://www.avirupbasu.com`, holds some of his work and one may get in contact with him there. Being a part-time blogger, he blogs about topics he is interested in. Currently, he is working on autonomous robot control using SONAR and GPS. He dreams about doing research and development in his areas of interest.

Roberto Gallea, PhD, has been a computer science researcher since 2007 at the University of Palermo, Italy. He is committed to researching fields such as medical imaging, multimedia, and computer vision. In 2012, he started enhancing his academic and personal projects with the use of analog and digital electronics and with a particular involvement in the open source hardware and software platforms, such as the Arduino. Besides his academic interests, he conducts personal projects, which are aimed at producing handcrafted items such as musical instruments, furniture, and LED devices using embedded invisible electronics. He also collaborates with contemporary dance companies on digital scenes and costume designing.

Vincent Gijsen is an all-round type of a guy. With a bachelor's degree in embedded systems, a masters in information science, work experience in a Big Data start-up, and being currently active as a security officer and cyber security consultant in industrial and infrastructure environments, he has a broad range of interests. In his spare time, he likes to fiddle with lasers, microcontrollers, and other related electronics.

I would like to thank Packt Publishing for their pleasant cooperation and their ability always present interesting reads to review like: Storm Blueprints: Patterns for Distributed Real-Time Computation, and Arduino Development Cookbook as well as my girlfriend: Lisa-Anne, for her support.

Randy Schur is a graduate student in mechanical engineering at the George Washington University. He has experience working with the Arduino, robotics, and rapid prototyping, and has worked on the book *Arduino Computer Vision Programming* by Packt Publishing Pvt. Ltd.

Fangzhou Xia is currently pursuing a master's degree in mechanical engineering (ME) at the Massachusetts Institute of Technology (MIT). He received his bachelor's degree in ME from the University of Michigan (UM) and a bachelor's degree in electrical and computer engineering at Shanghai Jiao Tong University (SJTU). His areas of interest in mechanical engineering include system control, robotics, product design, and manufacturing automation. His areas of interest in electrical engineering include Web application development, embedded system implementation, and data acquisition system setup.

www.PacktPub.com

Support files, eBooks, discount offers, and more

For support files and downloads related to your book, please visit www.PacktPub.com.

Did you know that Packt offers eBook versions of every book published, with PDF and ePub files available? You can upgrade to the eBook version at www.PacktPub.com and as a print book customer, you are entitled to a discount on the eBook copy. Get in touch with us at service@packtpub.com for more details.

At www.PacktPub.com, you can also read a collection of free technical articles, sign up for a range of free newsletters and receive exclusive discounts and offers on Packt books and eBooks.

https://www2.packtpub.com/books/subscription/packtlib

Do you need instant solutions to your IT questions? PacktLib is Packt's online digital book library. Here, you can search, access, and read Packt's entire library of books.

Why subscribe?

- Fully searchable across every book published by Packt
- Copy and paste, print, and bookmark content
- On demand and accessible via a web browser

Free access for Packt account holders

If you have an account with Packt at www.PacktPub.com, you can use this to access PacktLib today and view 9 entirely free books. Simply use your login credentials for immediate access.

Table of Contents

Preface

The Arduino platform makes it easy to get started with programming and electronics, but introducing wireless communication in your project can get complicated quickly. The XBee wireless platform hides most of the complicated details from you, and this book provides a step-by-step guide to using XBee modules with Arduino.

This book describes an example wireless sensor network, and invites you to build that network yourself. By following the steps in each chapter, you will build a network that can measure temperature and humidity in various rooms of your house, collect that data online, and automatically control your heating and/or cooling system to maintain the proper temperature in your house. This temperature can be configured through an online dashboard, ultimately putting control back in your hands.

All the concepts needed to build this example network will be explained, so you will have the knowledge to build your own project using these same concepts. Concepts that are closely related, but beyond the scope of this book, will be mentioned and appropriate references will be given so you can find out more if needed.

What this book covers

Chapter 1, *A World without Wires*, introduces the XBee platform and shows how to use the XCTU program to control and configure XBee modules. It covers the AT and API modes, firmware updates, and ZigBee network creation and security. It also shows how to transmit your first messages between two XBee modules.

Chapter 2, Collecting Sensor Data, provides more details on wiring up XBee modules, and introduces the xbee-arduino library that lets an Arduino take control of an XBee module. Reading a sensor, designing a packet format, and wirelessly transmitting data are discussed; thus, by the end of this chapter you will have a basic wireless sensor network where one or more Arduinos read temperature and humidity data and this is wirelessly collected by another Arduino.

Chapter 3, Storing and Visualizing Your Data, covers storing and visualizing your collected data with the Beebotte cloud service, using an Internet-connected Arduino and the MQTT protocol. Storing and visualizing your data on your own computer, using a Python program and database, are also briefly discussed.

Chapter 4, Controlling the World, shows how to let your network control things in addition to monitoring them, such as heating and/or cooling your house. Simple on/ off control is covered in detail, either using a relay module connected to an Arduino, or using an off-the-shelf wireless power socket that supports the ZigBee Home Automation protocol.

Chapter 5, Standalone XBee Operation, lets you implement simple devices without using an Arduino, by letting the XBee module directly control or measure things. You will see how to simplify the relay module from *Chapter 4, Controlling the World* and add window open/closed detection to your network.

Chapter 6, Battery Power and Sleeping, discusses options for battery-powering your projects, as well as techniques to reduce the power used. This includes some hardware techniques, as well as applying various sleep modes to drastically reduce XBee and Arduino power usage when they are idle.

What you need for this book

To upload programs (sketches) to your Arduino boards, you will need the Arduino IDE, which can be downloaded from `http://www.arduino.cc` Version 1.6.5 was used in this book, but it is recommended to get the newest version. This book assumes you are already familiar with this program and know how to write and upload a sketch, which will not be covered in this book.

For configuring and interacting with XBee modules, you will need the XCTU program, which can be downloaded from `http://www.digi.com/xctu`. Version 6.2.0 was used in this book, but it is recommended to get the newest version. No prior experience with this program is needed; it will be introduced in detail.

The example sketches in this book use a number of Arduino libraries. These libraries are:

- AltSoftSerial by Paul Stoffregen (version 1.3.0)
- XBee-Arduino library by Andrew Rapp (version 0.6.0)
- DHT sensor library by Adafruit (version 1.2.0)
- Adafruit MQTT library by Adafruit (version 0.11.1)
- Adafruit SleepyDog library by Adafruit (1.0.0)
- Adafruit CC3000 library by Adafruit (version 1.0.3 optional)

In general, it is recommended you download the newest version of a library using the library manager in the Arduino IDE. However, if you run into problems because a library has made changes that are not backwards-compatible, it might be useful to test the exact same version of a library that was used when writing this book. These versions are shown above, but also included in the provided code bundle.

The introduction of each chapter indicates what hardware you will need for the examples in that chapter. This always lists the hardware required to build an example once, sometimes also sharing items between examples. If you want to expand your network to include multiple temperature sensors, window sensors, and so on, you will of course need multiples of some of the components listed.

Who this book is for

This book is for those who have been playing with the Arduino platform and now want to make their creations wireless. There is no need to be a veteran programmer, though a basic understanding of the Arduino and Arduino programming is assumed. All examples make use of ready-made hardware, so no soldering skills are required and, electronically, things are limited to just connecting a few pins and wires. This book will also prove useful for anyone wanting to integrate XBee with microcontroller platforms other than the Arduino, since most of the advice regarding XBee is not specific to the Arduino.

Conventions

In this book, you will find a number of text styles that distinguish between different kinds of information. Here are some examples of these styles and an explanation of their meaning.

Code words in text, folder names, filenames, file extensions, pathnames, dummy URLs, user input, and configuration values are shown as follows: "Sending a packet is handled by the sendPacket() function."

A block of code is set as follows:

```
// the loop function runs over and over again forever
void loop() {
  digitalWrite(13, HIGH);    // turn the LED on (HIGH is the voltage
level)
  delay(1000);               // wait for a second
  digitalWrite(13, LOW);     // turn the LED off by making the voltage
LOW
  delay(1000);               // wait for a second
}
```

Any command-line input or output is written as follows:

openssl rand -hex 16

New terms and **important words** are shown in bold. Words that you see on the screen, for example, in menus or dialog boxes, also appears in bold like this: "Click on the **Update Firmware** button to replace the firmware of your device."

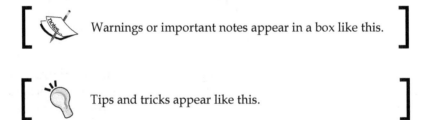

Warnings or important notes appear in a box like this.

Tips and tricks appear like this.

Reader feedback

Feedback from our readers is always welcome. Let us know what you think about this book—what you liked or disliked. Reader feedback is important for us as it helps us develop titles that you will really get the most out of.

To send us general feedback, simply e-mail feedback@packtpub.com, and mention the book's title in the subject of your message.

If there is a topic that you have expertise in and you are interested in either writing or contributing to a book, see our author guide at www.packtpub.com/authors.

Customer support

Now that you are the proud owner of a Packt book, we have a number of things to help you to get the most from your purchase.

Downloading the example code

You can download the example code files from your account at `http://www.packtpub.com` for all the Packt Publishing books you have purchased. If you purchased this book elsewhere, you can visit `http://www.packtpub.com/support` and register to have the files e-mailed directly to you.

Errata

Although we have taken every care to ensure the accuracy of our content, mistakes do happen. If you find a mistake in one of our books—maybe a mistake in the text or the code—we would be grateful if you could report this to us. By doing so, you can save other readers from frustration and help us improve subsequent versions of this book. If you find any errata, please report them by visiting `http://www.packtpub.com/submit-errata`, selecting your book, clicking on the **Errata Submission Form** link, and entering the details of your errata. Once your errata are verified, your submission will be accepted and the errata will be uploaded to our website or added to any list of existing errata under the Errata section of that title.

To view the previously submitted errata, go to `https://www.packtpub.com/books/content/support` and enter the name of the book in the search field. The required information will appear under the **Errata** section.

Piracy

Piracy of copyrighted material on the Internet is an ongoing problem across all media. At Packt, we take the protection of our copyright and licenses very seriously. If you come across any illegal copies of our works in any form on the Internet, please provide us with the location address or website name immediately so that we can pursue a remedy.

Please contact us at `copyright@packtpub.com` with a link to the suspected pirated material.

We appreciate your help in protecting our authors and our ability to bring you valuable content.

eBooks, discount offers, and more

Did you know that Packt offers eBook versions of every book published, with PDF and ePub files available? You can upgrade to the eBook version at www.PacktPub.com and as a print book customer, you are entitled to a discount on the eBook copy. Get in touch with us at customercare@packtpub.com for more details.

At www.PacktPub.com, you can also read a collection of free technical articles, sign up for a range of free newsletters, and receive exclusive discounts and offers on Packt books and eBooks.

Questions

If you have a problem with any aspect of this book, you can contact us at questions@packtpub.com, and we will do our best to address the problem.

1
A World without Wires

In this chapter, you are going to set up some XBee modules and send your first couple of bytes wirelessly. You will see how to wire up an XBee module, how to flash and configure it using the XCTU program, and how to manually transmit data. You will also learn about ZigBee networks, addressing, and network formation.

To follow the examples, the following components are recommended:

- Two XBee ZB modules (sometimes called "Series 2" modules, ordering code XB24-Z7xIT-xxx, https://www.sparkfun.com/products/11217).

- One SparkFun XBee Explorer USB adapter (https://www.sparkfun.com/products/11812).

- One SparkFun XBee shield (https://www.sparkfun.com/products/12847). Do not forget the stackable headers if they are not included with the shield.

- One Arduino Uno r3 (https://www.arduino.cc/en/Main/ArduinoBoardUno).

Together, they will look like this:

Note that the XBee ZB modules mentioned will likely be replaced by a newer version soon (ordering code XB24CZ7xIT-xxx). One of the modules shown in the preceding photograph is the new **S2C** module, the other is the older **S2** module.

The first two chapters include instructions on using other types of hardware too, so if the preceding hardware items are not easy to obtain, or you want to use different hardware for another reason (because you need features offered by other XBee modules, or you want to save a bit of money by using a second shield instead of the SparkFun Explorer board), read these two chapters first before you decide what to order.

If you run into problems at any time during this chapter, you could refer to *Chapter 2, Collecting Sensor Data*, which contains more details on how to connect XBee modules and has a troubleshooting section.

XBee radio hardware

For connecting multiple Arduino boards wirelessly, there are numerous add-on boards and modules available. Options range from simple low-power 433Mhz transmitters that offer raw radio access, and require everything from addressing and error correction to encryption to be done on the Arduino, to complex modules that take care of all radio processing, encryption, and mesh routing (and even include a programmable microcontroller).

This book will focus on the XBee modules manufactured by Digi International. These modules are easy to use, well supported by Arduino libraries, and not too expensive. Digi offers a number of different XBee product families, each with unique features and transmission range. Since all the XBee modules have a similar serial interface, configuration values, and hardware pinout, they are largely interchangeable. Advice and experience that apply to one often also apply to the others.

XBee product families

Here is an overview of all the XBee product families that Digi currently produces and that use the common XBee through-hole pinout (a few other families use a surface-mount design, but these are not discussed here):

- **XBee and XBee-PRO 802.15.4**: These modules let you use the radio protocol defined in the IEEE 802.15.4 standard directly. This is a fairly simple protocol designed for low-power, low-data-rate communication. It operates in the 2.4Ghz spectrum, just like Wi-Fi (defined in the IEEE 802.11bgn standards), but is a lot simpler, has a lower throughput, and has a much lower power usage. These modules support no (mesh) routing: only modules that can directly hear each other can communicate.

- **XBee and XBee-PRO ZB**: These modules use the ZigBee Pro radio protocol for communicating. ZigBee is a layer on top of 802.15.4 that adds mesh routing and extended network management capabilities. Mesh routing allows two modules to communicate even when they are not within range of each other, by letting other modules forward data on their behalf.

  On top of basic communication, ZigBee application profiles define standard messages and commands to allow, for example, remote-controllable lamps, switches, or other equipment from different manufacturers to interoperate.

- **XBee and XBee-PRO DigiMesh 2.4**: These modules run a proprietary mesh protocol developed by Digi, also in the 2.4Ghz spectrum. This protocol is comparable with ZigBee, but important differences are that all nodes are equal, can do routing, and even routing nodes can sleep to save power. These modules use the same **S1** hardware as the XBee 802.15.4 modules, so each can be converted to the other by replacing the firmware version.

- **XBee-PRO 900HP**: These modules run in the lower 900Mhz spectrum, giving them an extended range (up to 14km line-of-sight) and making them less sensitive to obstacles than the 2.4Ghz modules. They also use the proprietary DigiMesh protocol. Due to regulatory limitations, these modules are only usable in North America and a few other selected countries.

- **XBee-PRO 868**: These modules run in the lower 868Mhz spectrum, allowing up to 40km line-of-sight range. These modules do not use any meshing protocol; instead, they employ a proprietary protocol that only allows direct communication between modules. Due to regulatory limitations, these modules are only usable in Europe.

- **XBee Wi-Fi**: These modules use the common Wi-Fi protocol (802.11bgn). By default, these modules join the network of an existing access point, but they can also be configured to run as an access point, run an ad-hoc network, or use Wi-Fi Direct. Due to the nature of Wi-Fi, no mesh routing is supported and power usage is significantly higher than, for example, the XBee 802.15.4 and XBee ZB modules; however, throughput can also be a lot higher.

Sometimes you will also see the terms **Series 1**, which is the name originally used for the XBee 802.15.4 family, and **Series 2**, which is a retired family that used the same hardware as XBee ZB but with a different firmware (ZNet 2.5) and radio protocol. This old naming is still reflected in the names used for individual hardware boards, which use names such as S1, S2, S2B, and so on.

Versions and variants

As noted earlier, some modules are available in normal and PRO variants. These PRO modules use the same radio and serial protocol as the regular modules, but feature a more powerful transmitter and more sensitive receiver, allowing for a significantly extended range (and also requiring more power, of course). Since the radio protocol is the same, the normal and PRO modules of the same family can be intermixed.

Most of the XBee modules are available in a few different versions, differing only in the antennae they use. The easiest are the PCB antenna, which uses copper on the circuit board as the antenna, and the wire antenna, where a short piece of wire sticks up from the board. Both have similar performance. There are also modules available with a u.FL or RPSMA antenna connector, allowing them to have an external antenna (useful for projects inside a box, needing maximum reception). When in doubt, get the PCB antenna version, as it is the least fragile and performs well.

For a few more details about the available boards, also see this guide from SparkFun: `https://www.sparkfun.com/pages/xbee_guide`.

In the XBee ZB context, Pro can have two different and unrelated meanings, which can be confusing.

On the one hand, the XBee-PRO modules are more powerful versions of the regular modules. This is a distinction made by Digi in their hardware model names.

On the other hand, ZigBee Pro is the protocol used by the XBee ZB (and XBee-PRO ZB) modules. The latest version of the ZigBee specification, ZigBee-2007, defines two variants: Normal ZigBee and ZigBee Pro. ZigBee Pro is a bit more complicated, allowing for networks to scale to thousands of devices. Compatibility between these variants is limited; for this reason, most devices implement ZigBee Pro and the "normal" ZigBee protocol is not used very much.

In this book, all examples (and most of the discussion) focus on the XBee ZB modules, but most of the material presented also applies to the other modules. Where significant differences occur, these will be noted and (concise) instructions for other module types will be given. Therefore, if you decide to use the XBee 802.15.4, XBee DigiMesh 2.4, or XBee-PRO 868 modules (or any of the PRO variants), you can still work through this book normally. The XBee-PRO 900HP modules are not discussed in this book, but these are expected to work in a very similar way to the DigiMesh modules.

Instructions for the Wi-Fi modules are not included, since the setup and operation of Wi-Fi are significantly different from the other modules. However, a lot of the more general information (API frame format, XCTU operation, and AT commands) applies equally to these modules.

Official XBee documentation

For each of the XBee module series, Digi publishes a lot of documentation. This documentation can be accessed from the Digi website, through the **Support** section. The most important document is the **Product Manual**, which contains info on the hardware, network setup, a full list of commands and API frames supported, and so on.

Even though all the information you need for these examples is included in this book, you are encouraged to get familiar with the product manual for your XBee modules as well. The manual will be a lot more detailed in some areas and is easy to use as reference material for commands, configuration values, frame types, and so on.

Note that, within the ZigBee ZB family, there is a division between older boards (using S2 or S2B hardware) and newer boards (using S2C hardware). These boards are largely compatible, but there are two separate product manuals available with largely identical content (which are both titled *Product Manual: XBee / XBee-PRO ZigBee RF Modules*). Be sure to get the right one for your boards.

Your first transmission

In this section, you will send your first messages between two XBee modules. The first module will be attached to your computer through the XBee Explorer USB board, using the XCTU program to talk to it. To remove the need for a second Explorer board just for this example, the second module will be connected to an Arduino, which will simply print all data received through the serial monitor.

To let your XBee modules start a network, you will need to have exactly one **coordinator** module, and one or more **router** modules (the difference between these will be discussed later). XBee modules will usually default to being a router, so you will need to switch one into coordinator mode, as shown next.

This chapter will show brief instructions on connecting things with the recommended hardware. It is possible to use different hardware too, but it is recommended you read on until the first part of the next chapter, since that contains more details on how the connection actually works and what you should be aware of when figuring out how to connect your hardware.

Using the SparkFun XBee Explorer USB

This is by far the easiest way to connect an XBee module: just plug in the XBee module and connect the USB cable:

Be sure to disconnect the power (USB cable) when changing connections, or (un)plugging XBee modules into adapters and shields, to avoid damaging them.

You might have to install drivers for the FTDI USB-to-serial chip used by the Explorer, though this should have been taken care of if you have already installed the Arduino IDE (the older Arduino boards use the same chip and drivers). See the SparkFun site for more information on using the Explorer boards (`https://learn.sparkfun.com/tutorials/exploring-xbees-and-xctu`) or installing the FTDI drivers (`https://learn.sparkfun.com/tutorials/how-to-install-ftdi-drivers`).

If the drivers are installed correctly, a new (virtual) serial port will be made available; you can select it in the XCTU program later.

Getting and running XCTU

XBee configuration and test utility (XCTU) can be downloaded from `http://www.digi.com/xctu`. In this book, the Next Generation XCTU, version 6.2.0, was used, but using the newest version available is recommended. Starting with this 6.2.0 version, Linux is supported in addition to Windows and OS X.

After running through the installer and starting XCTU, the first step is to add a module that tells XCTU which serial port to connect to, and what settings to use. There are two buttons for this:

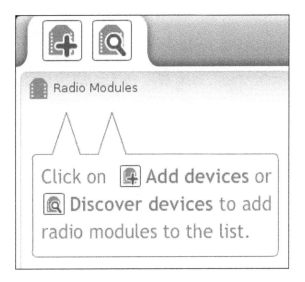

You will normally use the left **Add devices** button, which asks you for the settings to use. The right **Discover devices** button can be used when you are not sure about the right settings to use. Though it is not as magical as it sounds (you still need to select the serial port to use and select what different settings to try), XCTU will simply try all settings one by one until it finds one that works.

Go ahead and click on the left button. You will be greeted by a window where you can select the serial port and enter settings:

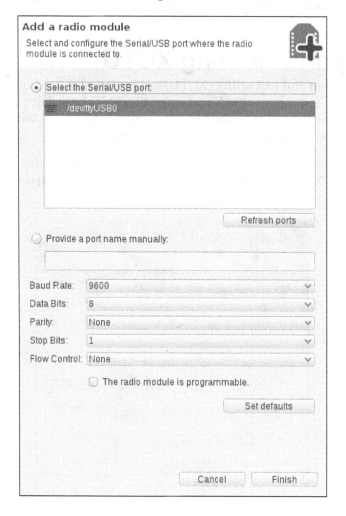

The serial port you need to use depends a bit on the platform. On Windows, it is likely the highest COM port listed (described as USB Serial Port). On OS X, it will look like usbserial–xxxxxxxx and, on Linux, it will be ttyUSBx. Unplugging your Explorer and refreshing the list is an easy way to figure out what port belongs to it.

By default, XBee modules are configured for 9600 baud, 8N1, so you can probably just leave all settings at their default, as shown in the preceding screenshot, and click on **Finish** right away. XCTU will open up the serial port and start talking to your XBee module to find out its type and firmware version. Shortly afterwards, the left side of the screen should show your module's information. If you click it, XCTU will further query your module for its current configuration settings. You should have something that looks somewhat like this:

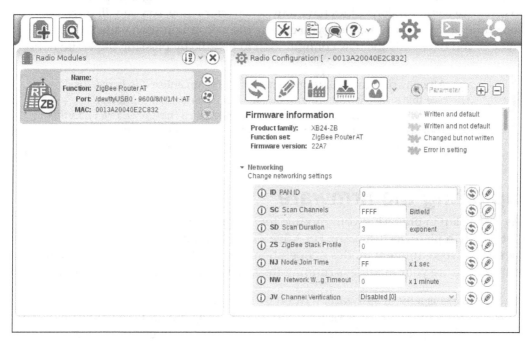

On the right part of the screen, you see what is called the **Configuration working mode** by XCTU. It shows all the current configuration values and lets you change them. If you change any of the values, they will not be sent to the device until you tell XCTU to write the changes. You can write all changed settings at once by clicking on the **Write radio settings** button at the top:

Alternatively, you can write just a single setting by clicking on the **Write** button next to the setting.

Note that both buttons update the current value in the module and let the module write the values to its non-volatile memory (so they are remembered across a power cycle).

In the rest of this book, when a configuration value is mentioned, this is the place to go to. All configuration values have a two-character name, so you will often see something like Set EE=1, meaning you change the EE configuration value to the value 1, and then write the new configuration to module by pressing the **Write** button.

 All numerical configuration values in XCTU are entered in *hexadecimal* (see this article if you are unfamiliar with hexadecimal numbers: https://learn.sparkfun.com/tutorials/hexadecimal). In this book, hexadecimal numbers are usually shown using the 0x prefix for clarity, but that prefix must be removed when entering values in XCTU.

Updating the firmware

As a first step in preparing your module for operation, you should check, and potentially replace, its firmware. For XBee ZB S2/S2B modules, this is needed to switch between different modes of operation, but for other modules it also makes sense to ensure your module is running the latest available firmware version.

The firmware is a bit of software, written by the manufacturer, that controls the XBee modules internally. It is responsible for operating the radio, taking care of the 802.15.4, ZigBee, and/or DigiMesh protocol processing (as well as any encryption), and communicating the results on the serial port.

For most XBee modules, there are several variants of the firmware, each with a slightly different **function set**. Some of these provide different features, but others are intended for hardware only. For now, you should just use the function set indicated next, but by the end of this chapter you will know what most of the function sets listed mean exactly.

To replace the firmware, some special commands have to be sent through the serial port to start the boot loader, which is a small piece of low-level software running on the XBee module that takes care of firmware updates. After starting the boot loader, the firmware file can be transferred through the serial port. Fortunately, there is no need to take care of this manually — the XCTU program can mostly handle firmware updates automatically.

To initiate a firmware update, select the module on the left, and click on the **Update firmware** button:

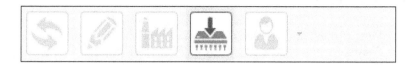

A window pops up, allowing you to select the **Product family**, **Function set**, and **Firmware version**. The right product family should already be selected (take a note of which one is selected in case you need to do recovery later). The first module will be the coordinator, so select the **ZigBee Coordinator AT** function set when using an XBee ZB S2 or S2B module, or **ZIGBEE Reg** function set when using S2C (where Reg is for *Regular* and indicates that this is not XBee-PRO). Finally, select the highest version listed to use the most recent firmware version.

You should probably also uncheck the **Force the module to maintain its current configuration** checkbox, in order to ensure you start with a clean factory default configuration.

Your selection should look like this:

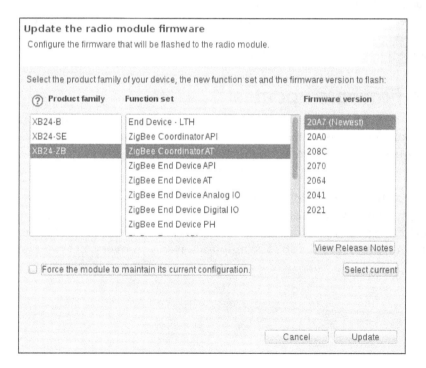

By clicking on **Update**, the XCTU program will start replacing the current firmware with the one selected. This takes around half a minute to complete; be sure not to unplug the board during the process.

 Version numbers for XBee modules are a bit confusing at first glance. They consist of four *hexadecimal* digits. Of these digits, the second digit is not really part of the version number, but indicates the function set used. For example, at the time of writing, 21A7 is the most recent **ZigBee Coordinator API** version, while 23A7 is the most recent **ZigBee Router API** version. This might suggest that the router API firmware is more recent, but both of these are equally recent and based on the same codebase; they just use a different function set.

Furthermore, the third digit indicates whether this is a stable (even) or beta (odd) version. The previously listed versions are stable, since their third digit is even (0xA = 10).

Note that, when you replace an existing coordinator's firmware, it will also start a new network. If other XBee modules previously joined the coordinator's network, it is possible that these are now joined to the old network and will no longer talk with the new coordinator. You can check by looking at the OI configuration value, these should be equal on all modules. The easiest way to force the router to leave the old network, is to change its ID configuration value, click the **Write** button, then change ID back to its previous value and click **Write** again. Doing this on the router(s) after you change the coordinator's firmware should prevent issues.

Failed firmware updates

Replacing the firmware is not entirely without risk. If the process is interrupted, or fails for some other reason, there is a chance that the module will end up without a working firmware. This prevents the module from working normally, but also prevents a subsequent firmware update from being started in the normal way.

If the regular firmware is broken, recovery requires connecting the XBee to your computer with hardware flow control (RTS/CTS) lines connected. The SparkFun Explorer USB board supports this out-of-the-box, but see this page for information on using an Arduino Leonardo for configuration and recovery: http://www.stderr.nl/Blog/Hardware/Electronics/Arduino/XBee/XctuAndArduino.html.

For exhaustive instructions on this recovery procedure, refer to the Digi knowledge base (search for XBee recovery).

Configuration

For this initial example, only minimal configuration is needed:

- The first module must be set to be a coordinator. When using the XBee ZB S2 or S2B modules, this is already done by uploading the coordinator firmware. When using the S2C modules, you will have to set CE=1 (**Coordinator Enable**) to turn the first module into a coordinator.

- The destination address for transmissions must be configured. This is determined by the DH and DL (Destination High and Low) configuration values. Together, they contain a 64-bit address of the destination (read on for more info on these addresses). For this example, it is best to set DH=0 and DL=0xFFFF, causing messages to be broadcast to all other nodes.

For example, to change the destination address:

1. Make sure you are in the **Configuration working mode** using the button at the top right.
2. Scroll down the list of configuration values in the right part of the screen, until you find the **Addressing** section.
3. Set DH to the value 0 and DL to FFFF (leaving out the 0x part).
4. Finally, click on the **Write radio settings** button above the list to apply the changes.

Setting CE is similar, but the value can be found in the **Networking** section instead.

Talking to the XBee module

Now, to talk to your XBee module, you can use the serial console built into XCTU (using another serial program is also possible, but do not forget to disconnect in XCTU first). To do so, head over to the **Consoles working mode** by clicking on its button at the top right:

Then click on the **Open connection** button in the top left of this area to enable the console:

The console is divided into two parts horizontally. The left side shows all data sent (blue) or received (red) as text and lets you type text to be sent. The right shows the same, but shows each byte value in hexadecimal.

Go ahead and type something in the left area. Anything you type there will be sent to the XBee module, which will then directly send the data through the radio again. So, you have sent your first radio message now, even though there is nobody listening yet.

To allow configuring the module, a special **command mode** is supported too (this is what is used under water when you change a configuration value in XCTU). By default, this command mode can be entered by sending a special sequence to the XBee module: a pause of at least 1 second, three iterations of a plus sign (**+++**, no Enter), and another pause of at least 1 second. When command mode is entered, the XBee module will reply with **OK**, and commands can be sent to the XBee module, terminated by a newline. These commands all start with AT, inspired by the **Hayes command set** originally designed for dial-up modems. Since these commands use normal text, it is easy to interact through AT mode simply by using a serial terminal. When no command is given for a while (10 seconds by default), the module leaves command mode again.

Try entering command mode and giving the **ATVR** command (do not forget to press Enter), which should retrieve the current firmware version:

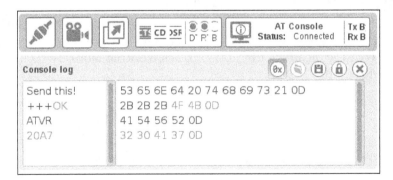

Receiving data

Now you have set up an XBee module to transmit data, the next step is to receive this data again on another module. The default firmware and configuration are probably sufficient for this but if they do not work, connect the second module to XCTU using your Explorer board and update the firmware to the **ZigBee Router AT** version (XBee ZB S2/S2B only) and check the DH and DL values.

To set up your second XBee module, you will use an Arduino Uno and the SparkFun XBee shield to forward all received data to the computer. Again, if you want to use different hardware, skip ahead to *Chapter 2, Collecting Sensor Data*, which has more details on how to connect an XBee module to an Arduino and how to modify the sketch to support a different setup.

Uploading the sketch

The sketch used for receiving data is available in the code bundle as SerialDump. ino. The exact contents of the sketch will not be discussed here, though you are encouraged to look inside to see what happens. Basically, the sketch opens up a serial connection to the XBee module at 9,600 bps (bits per second), and a serial connection to the computer at 115,200 bps. Any data received from the XBee module is forwarded to the computer, both in hexadecimal and readable *ASCII* format.

This sketch uses the AltSoftSerial library, so be sure to install that through the library manager, or download it from https://www.pjrc.com/teensy/td_libs_ AltSoftSerial.html.

 It is recommended you upload the sketch before connecting the XBee module, to prevent damage caused by the pins being in an unexpected state (this can cause a short-circuit in some cases).

Connecting the XBee

To connect the XBee module to your Arduino, you will use the SparkFun XBee shield. To allow talking both to the computer and the XBee at the same time reliably, you will need some trickery to connect the XBee's serial pins to AltSoftSerial's pins 8 and 9. This will be explained in more detail in *Chapter 2, Connecting Sensor Data*, but for now:

1. Put the switch on the shield into the DLINE position.
2. Connect pin 2 to pin 8 and pin 3 to pin 9 on the shield.
3. Insert the XBee module.

This should look a bit like this:

Receiving data

To actually receive data, open up the serial monitor in the Arduino IDE and configure it for 115200 baud. Now head back to the console in XCTU and type a message to be sent:

It is best to first type the message in a text editor, such as Notepad, and then paste it into XCTU. This makes sure the entire message is sent in a single radio packet. If you type in XCTU directly, each letter will be put into its own packet, which can cause delays and, when broadcasting, even mess up the ordering of letters in some circumstances.

In the serial monitor, you should see the same message being received:

```
Starting....
48 65 6C 6C 6F 2C 20 77   Hello, w
6F 72 6C 64 21            orld!
```

As you can see, any data sent to the XBee module on the sending side is reproduced byte-for-byte identically on the receiving side. For this reason, this mode of operation is commonly called **transparent mode** and can be useful to quickly make an existing serial connection wireless.

The sketch also allows typing a message into the serial monitor, which will be sent to the XBee module and then on to the other XBee module, and is shown in XCTU, so go ahead and try that too.

Switching to API mode

Until now, you have seen the XBee module work in **AT mode** or **transparent mode**. Any data the XBee module receives on its serial port is transmitted as it is by the radio, and any data received by the radio is forwarded through the serial port directly.

AT mode can be useful for simple point-to-point radio links, or when talking to existing serial devices (where you have no control over the serial protocol used). Transparent mode can also be used to upload sketches to a remote Arduino through the air (not covered in this book). Most examples and tutorials available on the Internet use this transparent mode for communicating.

However, AT mode is also limited. Data can only be sent to a single preconfigured address. When receiving data, there is no indication as to the source of a received byte. Furthermore, when receiving data from multiple nodes, their messages might end up being interleaved if no special care is taken. Finally, AT mode does not support sending more advanced ZigBee messages, which are needed to interoperate with some existing ZigBee devices—for example, using the ZigBee Home Automation protocol.

To alleviate these limitations, the XBee modules support **API (Application Programming Interface)** mode. In this mode, all serial communication uses a binary protocol, consisting of **API frames** (also called **API packets**). Each of these frames starts with the 0x7E byte to indicate the start of the frame and contains a frame length, a checksum, a frame type, and extra type-specific data.

0	1	2	3	4 ... n	n + 1
Frame Start 0x7E	Frame length MSB	 LSB	Frame type	Frame Data	Checksum

Examples of frames that can be sent to an XBee module are **Transmit Request** and **AT Command**; examples of frames sent by the XBee module are **Receive packet, AT Command Response**, or **Modem status**.

Each of these frame types has its own structure, containing a number of parameters that indicate the operation that should take place, or the event that occurred. If you look at the **Transmit request** frame, for example, this contains a field for the destination address, some transmission options, and the actual data to send.

Because of this format, it is easy to send data to different nodes, without having to change any configuration. Since data received is similarly wrapped in frames, you will know about the source of every message received too, which allows you to build more complex networks.

Because API mode uses a binary frame format, it is a bit harder to talk to the XBee module manually (you cannot just open up a serial console and start typing like you did before). However, for configuration and testing, the XCTU utility can take care of the frame generation and parsing for you; on the Arduino, there is a good library to do this.

For the XBee ZB S2/S2B modules, there is a separate firmware for AT mode and for API mode. To switch from AT mode to API mode, or vice versa, you have to replace the firmware. For the other XBee modules (including the XBee ZB S2C), a single firmware can run in both modes. Using the default setting of AP=0, they run in AT mode; however, by configuring AP=1 or AP=2, they run in API mode. There are two variants of API mode: with (AP=2) and without (AP=1) escaping some special characters (see the product manual for how exactly this escaping works).

In the rest of this book, all XBee modules will be configured for API mode with escaping enabled (AP=2), for maximum flexibility and robustness for your sensor network (note that the Arduino library used currently requires the use of escaping too). AT mode will not be covered in this book beyond what you have seen so far, but check out the product manual and resources online for more info on AT mode if you need it.

First module in API mode

To take advantage of API mode, you will need to configure your modules to use it. First, try switching your coordinator to API mode:

- For XBee S2/S2B modules, replace its firmware with the **ZigBee Coordinator API** version, as shown earlier. This step is not needed for other modules.

  Remember to change ID and change it back again on the other XBee module when you replace the coordinator's firmware, to force the other module to join the new network.

- Set AP=2 to enable escaping. XCTU will detect that escaping mode is now enabled and automatically applies the escaping to all data it sends and receives.

With this change, the XBee module will no longer allow typing of messages directly into the serial console, but expects to receive properly formatted API frames instead.

Sending data

To send these frames, head over to the **Consoles working mode** at the top-right again. Click on **Open connection** like before to connect to your XBee module.

You will notice that the window looks different from before. The screen is still divided into two parts horizontally, where the left part (directly under **Frames log**) will show all API frames sent to or received from the XBee module. This includes any frames sent by XCTU automatically (for example, when changing configuration parameters or doing a network scan), any replies to frames sent, as well as any data packets received by the radio. The right part shows details for any selected API frame.

Right now, this area should be empty, but you are going to change that by telling the XBee module to send out a message again. To do so, you will create a **ZigBee Transmit Request** API frame using XCTU and send it to the XBee module. At the bottom of the window, there is a **Send frames** section, which can be used to send API frames to the module. To prepare an API frame to send, click the **Add frame** button:

In the first input box, you can input a descriptive name for the frame. In the second input box, you can enter the raw bytes of the API frame. You could dig through the product manual to figure out how to construct a **ZigBee Transmit Request** API frame manually (which is still a good exercise), but it is easier to use the **Frames Generator** tool included with XCTU. You can access this tool by clicking the button at the bottom of the window:

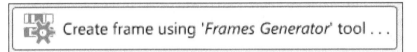

In the Frames Generator tool, leave the **Protocol** value at the default, which should be appropriate for your module. The **Frame type** drop-down list shows all frame types defined for your particular device and protocol. Note that it also includes frame types that are normally only sent by the module, such as **Receive packet**. These are included for completeness, but will be ignored by the module when you send them to the module.

Downloading the example code

You can download the example code files from your account at http://www.packtpub.com for all the Packt Publishing books you have purchased. If you purchased this book elsewhere, you can visit http://www.packtpub.com/support and register to have the files e-mailed directly to you.

The frame you need looks like this:

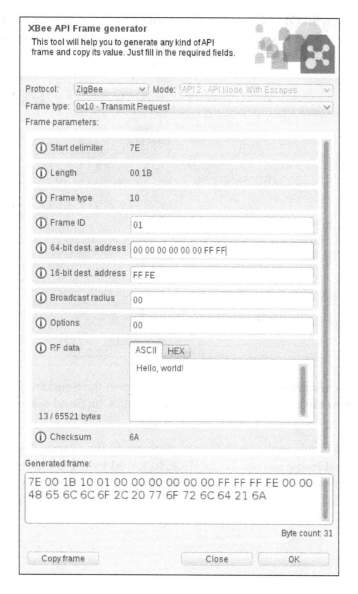

This selects the **Transmit request** frame type (called **ZigBee Transmit Request** in the product manual). Most of the other parameters are fine at their defaults. Under **64-bit dest. address**, enter 00 00 00 00 00 00 FF FF. This is the same broadcast address you configured in DH and DL before. When using API mode, these configuration values are unused; instead, the target address is specified for each outgoing frame individually.

Under **RF Data**, you can enter the actual message data to be sent.

After confirming with **Ok**, XCTU now shows the raw bytes of your `Hello, world!` frame. If you look closely, you can discover the `0x7e` start byte, `0x001b` as the frame length, `0x10` as the frame type, and the destination addresses:

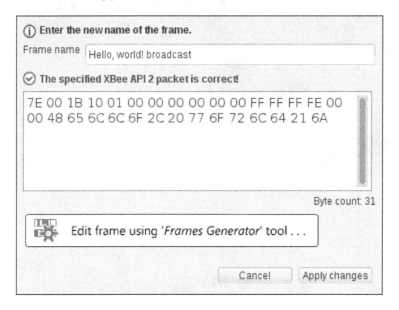

Click on **Apply changes** to make the frame show up in the **Send frames** list. To actually send it to the router module, select it and click on **Send selected frame** to the right.

Doing this immediately makes two frames show up in the **Frames log**. The first is the **Transmit Request** frame sent to the module, followed by a **Transmit Status** reply. Clicking the frames shows their details to the right. If everything went right, the status frame will show a **Delivery status** of **00 (Success)**, indicating that the frame was successfully sent.

Now, head back to the serial monitor that shows the output from your Arduino. This should show the "Hello, world!" message you transmitted:

```
Starting....
48 65 6C 6C 6F 2C 20 57   Hello, W
6F 72 6C 64 21            orld!
```

As you can see, you can exchange data between nodes running in AT and API mode just fine; each will just output the received data to the serial port in its own mode. You could try sending a bit of data back from the Arduino to the coordinator as well, to see how that shows up in XCTU.

Second module in API mode

To complete the transition to API mode, you should switch over the second XBee module, too. For this, plug it into the XBee Explorer module temporarily to replace the firmware if needed and set AP=2.

Then plug the second XBee module back into the Arduino shield and the first one back into the Explorer. If you use XCTU as before to let the first module send the "Hello, World!" message again, the Arduino serial monitor output will look different from before:

```
Starting....
7E 00 19 90 00 13 A2 00   ~.......
40 D8 5F 9D 00 00 02 48   @._....H
65 6C 6C 6F 2C 20 57 6F   ello, Wo
72 6C 64 21 3B            rld!;
```

Instead of printing just the message, a full API frame is printed. The message is contained inside, but also the API frame type (0x90), sender address (0x0013A20040D85F9D), message length (0x0019), and so on. Check out the section on API operation in the product manual to figure out exactly what each byte means.

If you try typing some data into the Arduino serial monitor now, you will see that this data no longer shows up in XCTU like before. The XBee module expects to receive a correct binary API frame, which is unlikely to happen if you just type some letters. In the next chapter, you will see how to use an Arduino library to properly format these API frames.

Starting and joining a network

To start sending data between your XBee modules, they all have to be joined to the same network first. In the previous examples, you have used a single coordinator and router, which together have automatically formed a network. In this section, you will learn a bit more about how these networks are formed and how addressing works.

The ZigBee protocol defines three different device types: coordinator, router, and end device. These are explained next:

- A **coordinator** starts a network and allows other devices to join it. It otherwise acts as a normal router. Each network must have exactly one coordinator.

- A **router** joins an existing network and will forward (**route**) packets on behalf of other members in the network, creating a mesh network. However, to maintain this routing functionality, routers cannot enter sleep mode and should usually be mains-powered.

- An **end device** is more limited: it cannot do routing itself, instead relying on a router (or the coordinator) to route its messages to their destination. Unlike routers, end devices can sleep to save power.

A typical network might look like this:

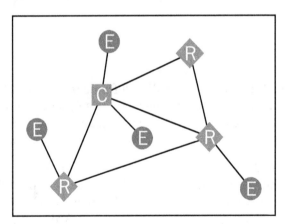

Note that the coordinator and routers can talk to each other (when they are close enough), but each end device can only talk to a single router (or the coordinator).

In the first chapters, you will use a single coordinator and one or more routers. In *Chapter 6, Finishing Touches*, end devices will be discussed in more detail.

In ZigBee, a network is identified by a **PAN** (**Personal Area Network**) identifier. When a coordinator is first started with its default settings, it will automatically start a new, unencrypted network by selecting a random PAN ID and a random channel. When a router or end device module starts, it will start scanning all channels for existing networks that allow new modules to join, select one, and join it. In the default configuration, coordinators and routers allow new devices to join the network without restriction.

There are a few related read-only **configuration** values supported, shown under the **Networking** category in XCTU:

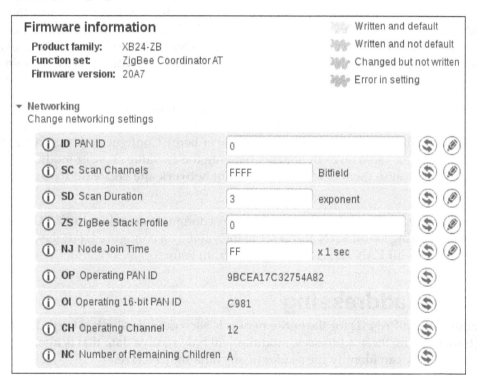

In the preceding screenshot, you can see the OP value. This is the **Operating PAN ID**: the PAN ID of the network created by the coordinator (or, on a router or end device, the PAN ID of the network successfully joined). The CH **Operating Channel** value shows the channel number selected by the coordinator (you might need to click on the **Refresh** button at the top to see the most recent values here).

Note that there is a second operating PAN ID here: OI. OP is a 64-bit long version, which is used in scan requests, beacons, and network joining, but OI is a 16-bit short version, which is used in most packets to save transmission time. If two different networks happen to have selected the same 16-bit PAN ID, the network can detect this and the conflict can be resolved by switching to a new 16-bit PAN ID. The 64-bit PAN ID will never change (unless of course the coordinator leaves the network, starts a new one, and ID is not set).

Also note the ID configuration value. This allows preconfiguring the 64-bit PAN ID on your modules by using a non-zero value. On a coordinator, this means the started network gets this 64-bit PAN ID. On routers and end devices, this means that the module will only join networks with a matching PAN ID and will ignore all others. Preconfiguring a PAN ID can help to prevent unexpected behavior (consider the case where your neighbor is also running a ZigBee network, and your nodes are joining his network instead of yours).

Go ahead and select some random 64-bit number for your network. This can be some easy to remember number, or whatever you like (unlike encryption keys, there is no security benefit of having a true random number here). Configure this number in the ID configuration value on both nodes. Changing the ID value on your modules will automatically cause them to leave their current network and create (or try to join) a new network.

You will see that, after this (and refreshing the configuration values in XCTU), the OP value (**Operating 64-bit PAN ID**) is set to the value you configured, but the OI value (**Operating 16-bit PAN ID**) is now different from before.

Module addressing

In addition to a PAN ID for the entire network, each module also has two addresses of its own. Using these addresses, modules will only receive data that is intended for them, and they can identify the sender of any data they receive.

The two different addresses for each module are:

- A **64-bit extended address**. This address is preconfigured in all XBee modules in the factory and can never be changed. These addresses are also called EUI-64 (Extended Unique Identifier). These identifiers are distributed by the IEEE, ensuring that every one of them is guaranteed to be globally unique (similar to MAC addresses in Ethernet adapters).

  The 64-bit address of your XBee modules can be found by looking at the SH and SL configuration values (which contain the upper and lower 32 bits of the address respectively), and is also printed on a sticker on the module itself (16 hexadecimal digits).

- A **16-bit short address**. This address is used in normal communications and is assigned to a module when it joins the network. Normally, this address does not change until the module leaves the network again, but there is a chance that two modules could end up with the same short address, in which case both modules will get a new short address assigned automatically.

  The coordinator always gets the 0x0000 short address, while 0xfffe is reserved for specifying an unknown address (this is usually used when the 64-bit address is known, but not the 16-bit address).

  The short address assigned to a module can be found by looking at the MY configuration value.

Network scanning and remote configuration

To verify that all devices have been connected to the same network properly, or to find out more about how they are connected, you can use the network scanning feature in XCTU. This feature can be accessed in two ways. The most extensive one is by switching to the **Network working mode** at the top right:

Start a network scan using the **Scan the radio module network** button:

Alternatively, you can also directly click on the **Discover radio nodes** button next to a connected module in the list on the left. The former method shows more details about the network, such as a graph containing links, and shows all devices found. The latter method shows detected nodes in a list and only shows XBee modules.

To help identify nodes, the **NI (Node Identifier)** configuration value in the **Addressing** section can be used. This is just a human-readable name for this XBee module, which can help you tell them apart (though it is not shown in the network scan with the 6.2.0 version of XCTU, unfortunately).

In addition to discovering remote devices, you can also configure any remote XBee modules, just as if you are connected directly to them. To do so, double-click the device in the graph, or select it in the discovery list, and click on **Add selected devices**. Go ahead and explore these XCTU features; they might come in handy later.

The Commissioning button and LEDs

Some XBee adapters or shields have a commissioning button, which can be used for some limited interaction with the module (such as briefly allowing new devices to join the network or doing a factory reset—see the product manual for details). Such a button simply connects to the AD0/DIO0 pin on one side and GND on the other, shorting them together when the button is pressed. If your setup does not have a button, it is easy to add one, or simply use a jumper wire to briefly make the connection to emulate a button press. Also, note that some adapters have a reset button (to reset the XBee and/or Arduino), so be sure to check what kind of button you have.

More commonly, adapters and shields have a LED connected to the XBee Associate/DIO5 pin (the associate LED) and/or the RSSI/DIO10 pin (the RSSI LED). The associate LED is lit continuously when a device is not joined to a network, but slowly blinks when the XBee module is successfully joined. It can also be used for more advanced diagnostics together with the commissioning button; see the product manual for details. The RSSI LED indicates the signal strength of the most recently received signal (the better the signal, the brighter the LED).

Both LEDs and the commissioning button can be disabled through the XBee configuration (to save power, or allow reusing the pin for a different purpose), but are enabled by default (though there will not always be a button or LED connected, of course).

Making your network secure

Using wireless communication in your project is very convenient, but it also opens up the possibility of others sniffing your data, or even sending packets into your network to influence your project. For this reason, securing your network with proper encryption methods is an important step that is best taken right away (since it is easy to keep postponing it otherwise).

Fortunately, the XBee modules make securing your network easy and handle all the details automatically. There are a few concepts you need to understand, though.

When you enable security in your XBee module (by setting the EE configuration value to 1), most packets being sent will be encrypted (preventing others from reading the messages) and protected by a message integrity code (preventing others from inserting messages into your network). Special network-management packets (such as beacons, join requests, and so on) are not encrypted, nor are the lowest level packet headers, but the actual data you send is always secured.

This encryption happens using a secret key, called the **network key** (NK). All members of the network must know this key to allow sending and receiving messages. This network key can be configured on the coordinator by setting the NK configuration value. The default value of 0 causes the coordinator to select a random key when it creates a network, which should be sufficient in most cases.

Distributing the network key

When a module joins the network, it automatically receives the network key from the so-called **trust center**. The trust center is responsible for distributing network keys throughout the network when a node joins, or when the key is changed.

In XBee networks, the trust center always runs inside the coordinator, but the ZigBee protocol itself makes provision for running the trust center in another module as well. The ZigBee specification also allows preconfiguring the network key on all nodes, but the XBee firmware does not currently support this.

By default (with the KY configuration value set to its default of 0), the trust center sends the network key to joining nodes without using any encryption. This obviously poses a security risk: if someone can spy on your network traffic while your modules are joining the network, they can see your network key. Even more, an attacker could simply have his own module join the network and the network key would be sent to him, unencrypted.

To reduce this risk, you can configure your network to not allow new nodes to join normally, and only enable joining briefly when you have a new node to add. The network will still be vulnerable during these brief periods, but if the network is not compromised while joining is enabled, it will remain secure while joining is disabled. Refer to the *Disabling network joining* section for more info on how to disable joining.

Trust center link key

To further secure the network and avoid having to send the network key unencrypted, ZigBee defines a second key, called the **trust center link key**. This key is manually configured in each module (using the KY configuration value) before it creates or joins the network. The trust center then uses this key to encrypt the network key when it is sent to new nodes joining the network. Eavesdroppers, not knowing the trust center link key, will not be able to decrypt the network key and so cannot read any of your data.

On XBee modules, there is a single trust center link key that must be known to all joining modules. The ZigBee protocol also allows for using a different trust center link key for each module, which would require more extensive configuration, but this is not supported by the current XBee ZB firmware.

There is a third kind of key in the ZigBee protocol: an application link key. One of these keys can exist for every pair of modules in the network and allows additional end-to-end encryption (on top of the existing network encryption). This encryption is referred to as **APS encryption** in the XBee documentation and allows two modules to privately communicate, without allowing intermediate routers or other modules that are joined into the network to decrypt their messages.

Currently, the XBee firmware does not support these application link keys, so they will not be discussed any further.

Trust center terminology differences

It turns out there is a bit of a discrepancy between the XBee documentation and the ZigBee protocol specification around the concept of the trust center.

According to the ZigBee specification, a trust center is always needed to distribute the network key to newly joining nodes (unless the network key is already preconfigured on all nodes, which the XBee modules do not currently support). Additionally, the trust center can be used to distribute application keys or propagate a network key change throughout the entire network.

In the XBee documentation, distributing the network key to newly joining nodes is assumed to be done by the coordinator, even if the trust center is not enabled (controlled through the EO configuration value). Enabling the trust center seems to be needed only to allow propagating network key changes.

Selecting encryption keys

When selecting a random key, be sure to always select a full-length key. It is possible to enter a shorter key in the KY and NK configuration values and a lot of the examples in the XBee documentation also do this. However, if you enter, for example, ABCD as a key, it will be prepended with zeroes, so the actual key will end up being 00000000 000000000000000000000ABCD. This severely limits the security—a four-digit key like this can be brute-forced in a matter of seconds using a regular computer.

A strong key should be properly random and be 32 (hexadecimal) digits long (which is 16 bytes or 128 bits). Keys such as these are effectively impossible to crack using brute force, even when using current supercomputers.

> If you have the OpenSSL software installed (the default on most Linux and OS X systems), you can easily generate a random key using the openssl command line utility:
>
> ```
> $ openssl rand -hex 16
> 077ef3b3e5121efa166445c63e18538d
> ```

Setting up your secure network

Now that you understand how security works in a ZigBee network, you can go ahead and restart your network with security enabled. To enable optimal security using a trust center link key, make the following configuration changes to both the coordinator and the router node (in either order):

- Configure EE=1
- Configure KY (the trust center link key) to a randomly selected key. All modules should use the same key here.

For example:

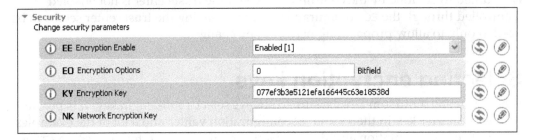

Note that the KY value can only be set, never read back, so be sure to store the key you set somewhere to allow adding new nodes to your network later.

Changing these values will automatically cause the modules to leave the current network and create a new network (coordinator) or find a new network to join (router).

After a few seconds, the new network should be formed. You can again transfer your Hello, world! message, but in a secure way this time.

Disabling network joining

By default, XBee modules allow new modules to join the network at any time. If you configured a trust center link key, joining is of course limited to nodes that know this secret key, preventing unauthorized access to your network.

However, as an additional layer of security (or, when not using a trust center link key, the primary layer), you can also disable all join attempts after you started your network and joined your devices. You can briefly re-enable joining later when you intend to add a new device to your network.

When a device wants to join a network, it searches for an existing router (or coordinator) that allows joining. Whether this is allowed depends on the NJ configuration value:

- When NJ is 0, joining is disabled
- When NJ is 0xff, joining is enabled
- When a router joins a network, and its NJ value is not 0 or 0xff, the router allows joining for NJ seconds
- When the NJ parameter is changed, joining is again enabled for that many seconds

- When the commissioning button is pressed twice quickly, joining is enabled for one minute

The NJ value is not a global value for the entire network, but each node has its own value and can decide on its own whether to allow new nodes to join. Even if the coordinator has joining disabled, the trust center will still send keys to nodes that joined through another router.

In general, it is a good idea to set NJ=0 on all modules by default, and then change the value on a nearby router or the coordinator (or use the commissioning button) whenever you need to add a node to the network.

Note that, for end devices, there is also a special rejoin procedure; refer to *Chapter 6, Finishing Touches* and the product manual for more information about this.

Other XBee families

The ZigBee network structure used by the XBee ZB modules is the most complex of all the XBee modules discussed in this book. The other modules do not require having a central coordinator or going through a join procedure, but instead rely on manually assigning a 16-bit network identifier (ID) and operating channel (CH) on all nodes (except XBee-PRO 868, which only has one channel available). Security is also simpler: every node has the network key (KY) manually configured. There is no key distribution mechanism.

For the XBee 802.15.4 family, this applies in the default configuration, but there is a second mode where a simple join procedure can allow nodes to automatically detect the right channel and/or network identifier to use (the network key must still be manually configured).

DigiMesh and XBee 868 networks use only a 64-bit address for each node and all transmissions. All nodes in an 802.15.4 network have a 64-bit address and can optionally have a manually assigned 16-bit address too (MY). Sending a packet in a 802.15.4 network uses either a 64-bit or 16-bit destination address (but, if you use a 64-bit address, the full address is transmitted over the air; there is no mechanism to look up the 16-bit address and use that in packets as with ZigBee).

XBee 802.15.4 and XBee-PRO 868 networks do not have any mesh capabilities, so you can only send messages between nodes that can directly hear each other. DigiMesh networks do support meshing similar to ZigBee.

Some other families also have device roles such as coordinator or end device, though the difference between them is less than in ZigBee networks. See the relevant product manual for details.

Configuration values

If you followed the preceding examples, you should have configured the following values on all your XBee modules:

- `AP=2` to enable API mode with escaping
- `ID=<your PAN ID>` to prevent joining other networks
- `EE=1` to enable security
- `KY=<random 32-digit key>` to prevent others without this key from joining the network
- `NJ=0` to prevent new nodes from joining

The changes to `DH` and `DL` were only needed for the AT mode examples and are no longer relevant in API mode.

Summary

In this chapter, you have learned about some wireless terminology and available hardware, and sent some bytes over the air. Your XBee network is even secured against eavesdropping. You have already used an Arduino as a simple forwarder of data, but it is time to put your Arduinos in control of the XBee modules and get them talking to each other! The next chapter will show how to connect your XBee modules to your Arduino and how to exchange some simple sensor data.

2
Collecting Sensor Data

In the previous chapter, you connected your XBee modules to your computer and sent a bit of data between them. In this chapter, you will take a few more steps and put your Arduino boards in control of the XBee modules. You will even send over some measured temperature and humidity data, creating your very first actual wireless sensor network.

You will build towards this step by step. The first step is to find out how to make an Arduino and XBee module talk to each other. Then you will send a simple `Hello, world!` message from one Arduino to another and finally replace that message with some actual live sensor data.

To follow the examples, the following components are recommended:

- Two XBee ZB modules (such as `https://www.sparkfun.com/products/11217`)
- Two SparkFun XBee shields (`https://www.sparkfun.com/products/12847`)
- Two Arduino Uno r3 boards (`https://www.arduino.cc/en/Main/ArduinoBoardUno`)
- A DHT22 (or DHT11) temperature and humidity sensor (`https://www.adafruit.com/products/385`)
- A 10kΩ resistor (optional)
- A breadboard (such as `https://www.adafruit.com/products/65`)
- Some jumper wires (such as `https://www.adafruit.com/products/153`)

If you want to deploy more than one sensor node, you will of course need additional components. As before, using different but similar components is possible. This chapter should provide insight into how things are connected so you can figure out how to use these other components too.

If you want to use another sensor (another temperature sensor, or even a completely different type of sensor), that is also perfectly possible. Most of the examples and code in this chapter will still apply, but you will need to figure out how to talk to your specific sensor (but Arduino libraries are available for most commonly used types).

Hardware setup

In this section, you will look in more detail at how to connect an XBee module. By the end of this section, you will better understand the connections described in the previous chapter, and will be able to figure out how to wire up other shields and adapters too.

Serial on XBee

Interfacing with an XBee module is not terribly complicated. All XBee modules offer a serial connection. In the most basic version, this involves just three pins on the XBee board: DIN (where the XBee module receives data), DOUT (where the XBee module sends data), and GND to establish a common ground reference.

This type of connection does not have any official name, but is commonly (but inaccurately) called **TTL** (**Transistor-to-Transistor Logic**) serial and is also used on the Arduino Uno between the integrated USB-to-serial converter and the main microcontroller.

It would be more accurate to say that it is an **asynchronous**, **single-ended** serial protocol running at 3.3V. Asynchronous indicates that there is no separate clock line for synchronization (both ends of the connection need their own clock), unlike synchronous protocols such as I2C or SPI, where a master exposes a clock for the slave(s) to use. Single-ended indicates a signal is carried on a single wire with voltage relative to a common ground, unlike differential protocols such as RS485 where two wires each carry a mirrored version of the same signal.

Another commonly used asynchronous serial protocol is RS232, which is used by normal serial ports on (mostly older) computers. It cannot be used to connect an XBee, because the higher voltages used would damage the XBee (see `https://learn.sparkfun.com/tutorials/serial-communication` for more info on TTL and RS232 serial).

Handshaking signals

In addition to the DIN and DOUT data pins, XBee boards also offer some additional handshaking signals: DTR, RTS, and CTS (originally defined in the RS232 specification). Of these, RTS and CTS can be used for **flow control**, preventing either side of the serial connection from sending data when the buffer on the other side is full. DTR is typically used to detect whether the serial port is opened or closed.

As mentioned before, these signals are also used for updating the XBee firmware and recovering from a failed update, but are not required for normal operation.

Since the Arduino serial libraries do not currently support any flow control, and most shields leave these pins disconnected, these signals will not be discussed in detail.

Voltage levels

XBee modules run at 3.3V, meaning their I/O lines also use that voltage. Directly connecting a 5V device (such as most Arduinos) to an XBee module is likely to damage it. All shields we have seen have some form of level conversion present on the DIN pins, but not all shields have level converters on the DOUT pins, since a 5V Arduino can read a 3.3V signal just fine. Some shields also connect the handshaking pins without any level converter.

However, if you accidentally set these pins to OUTPUT mode and write HIGH to them, the XBee can be damaged. Best to always double-check the pin connections and ensure that the right sketch is running before connecting an XBee module to an Arduino.

Serial on a computer

Normal computers do not typically offer any serial ports (older computers used to have RS232 ports, but usually no TTL serial connections). To solve this, people typically use some kind of USB-to-serial converter (one that uses TTL serial, not RS232 serial, such as https://www.sparkfun.com/products/9873). Make sure you use a 3.3V version, so you do not need any level converters. You will also need some way to connect the converter to the XBee board; using a breakout board (such as https://www.sparkfun.com/products/8276 with pins soldered on) allows using normal jumper wires or a breadboard for this.

To connect these generic converters, be sure to connect the XBee DOUT to the converter's RX (receive) pin and DIN to TX (transmit). The handshaking signals can be directly connected if the converter supports them.

There are a couple of XBee-specific converters, such as the SparkFun XBee Explorer USB used in the previous chapter, where you can just plug in the XBee board and a USB cable directly. The explorer also has all handshaking signals connected, making it ideal for firmware updates and recovery, too.

You might think you can just use an Arduino and XBee shield you already have to connect your XBee module to XCTU, but this turns out to be harder than it looks or even impossible, depending on the hardware used. See this post for a detailed write-up on the options and problems involved: `http://www.stderr.nl/Blog/ Hardware/Electronics/Arduino/XBee/XctuAndArduino.html`.

Serial on Arduino

To connect an XBee module to an Arduino board, the DIN and DOUT pins of the XBee must be connected to a serial port on the Arduino. There are two kinds of serial ports available:

- A **hardware serial port** (often called a **UART**, for Universal Asynchronous Receiver and Transmitter). This is a piece of hardware inside the microcontroller that can operate a serial line by itself, only needing attention from the microcontroller when a full byte was received or sent.

- A **software serial port**. This is a serial port implemented in software, using normal digital I/O pins. This requires the microcontroller to control and monitor the state of the I/O pins for each bit separately.

Included with Arduino is the SoftwareSerial library, which is not recommended. When any data is being transmitted, SoftwareSerial is unable to also receive data, which is likely to cause problems when communicating with the XBee module.

An alternative third-party library is AltSoftSerial, which uses interrupts and timers and is a lot more reliable. The downside is that it can only be used on two specific pins, possibly making wiring things up a bit more tricky.

Whenever possible, you should use a hardware serial port, because those are a lot more reliable and efficient. Software serial ports require a lot more processing time and are more sensitive to timing problems when other code and interrupts are involved.

However, a lot of Arduino boards (including the commonly used Arduino Uno) only have a single hardware serial port, which is also used to talk to the computer (through the USB-to-serial adapter integrated on most Arduino boards). If you connect the XBee module to this port and also open the USB serial port, these will conflict.

So, when using this hardware serial port for XBee, you cannot use it for communicating with the PC, or providing debug output. It also means that, while the XBee module is connected, you cannot upload new programs either (requiring removing the XBee board, or disconnecting the serial pins using any switch or jumpers present on the shield). On Arduino boards with a single hardware serial port, using a software serial port is usually the best option.

If you want to use a hardware serial port instead, you will need to use a board that has more than one serial port available. Some of these are:

- The Arduino Leonardo. The microcontroller used on this board connects directly to USB, leaving its single hardware serial port available to connect to an XBee module. For more info on the Leonardo, see: `https://www.arduino.cc/en/Main/ArduinoBoardLeonardo`.

- The SparkFun Fio v3. This board uses the same microcontroller as the Leonardo and even has an XBee socket built-in (no shield needed). Since it is not made by Arduino but SparkFun, you'll need a bit more work for the Arduino IDE to recognize this board and, since it does not have the standard Arduino board layout (it is smaller), you cannot use standard Arduino shields with it. Also note that the earlier Fio versions do not have this extra serial port available. For more info on the Fio v3, see:

  `https://learn.sparkfun.com/tutorials/pro-micro--fio-v3-hookup-guide?_ga=1.57187369.2059863491.1441280898#hardware-overview-fio-v3`.

- The Arduino Mega. This board has more resources, including four hardware serial ports, but is also bigger and more expensive. For more info on the Mega, see `https://www.arduino.cc/en/Main/ArduinoBoardMega2560`

Some third parties also have some suitable boards available, often based on the Atmega 1284P chip. Worth mentioning are the boards from SODAQ (`http://shop.sodaq.com`), which additionally have an XBee socket built-in and are optimized for low-power battery or solar-powered applications, making their boards interesting for setting up a sensor network.

The examples in this book have been mostly developed using an Arduino Uno, since these are easily available and flexible. The examples use the AltSoftSerial library for talking to the XBee module, which should be sufficiently reliable in most cases. All the example code is aimed at this setup, but the code is written in a way that allows using a different setup with minimal changes.

XBee shields

Once you have decided what serial port to use, you will need to connect the XBee modules to the corresponding pins (typically pins 0 and 1 for the hardware serial port, or pins 8 and 9 for AltSoftSerial, though this depends on the board). Connect Arduino RX to XBee DOUT and Arduino TX to XBee DIN.

On most shields, connecting the XBee to pins 0 and 1 is easy, but connecting it to pins 8 and 9 (used by AltSoftSerial on the Arduino Uno), can be more tricky. For example, when using the recommended SparkFun XBee shield, there is a switch to choose between pin 0/1 and 2/3, but there is no built-in way to connect the XBee serial pins to 8/9. Fortunately, this can be easily solved by setting the switch to **DLINE** (connecting the XBee serial pins to 2/3) and adding jumper wires to connect 2/3 to 8/9, as you have seen in the previous chapter.

A downside of this approach is that pins 2 and 3 cannot be used for anything else (and since those pins have interrupts available, a lot of shields use them). With a bit of modification on the XBee shield (cutting the traces to pin 2 and 3 of the Arduino, and soldering in some wires instead of using jumper wires in the pin header), this can be fixed if you need pin 2 or 3 for something else.

Other shields

There are some more XBee shields, available from various vendors, that can also be used. Other shields might need similar tricks, or might be easier (the ITEAD and Seeed Studio XBee shields allow connecting DOUT and DIN to any I/O pin using a long row of jumper pins, for example). Some shields have the connection fixed to pin 0 and 1 (such as the official Arduino Wireless shield), which will be trickier when you want to use a software serial port.

When looking at alternative shields, be sure to check the documentation and schematics carefully so you know how things are connected. There are different-level conversion methods being used and some shields have a way for the XBee to reset the Arduino (which you do not want to trigger accidentally, of course). As mentioned before, some shields connect the handshaking signals without any level conversion, so make sure those pins are not used for anything else.

Software setup

Now that the hardware is covered, it is time to look at the software side of things. In this section, you will create a small sketch (called Connect.ino in the code bundle) that makes the Arduino send a VR command to the XBee module to query the current firmware version of the XBee module. The Arduino code is set up so that all API frames received from the XBee module are printed, so you can see the reply to this command as well, confirming that connectivity is working. No data is transmitted wirelessly yet; this just tests the Arduino-to-XBee connection.

Example sketches

In this and the following chapters, you will write some Arduino sketch code. All of the example code shown in the book is also available in the code bundle, which you can download from the Packt website.

The example code in this book is sometimes a bit more verbose than strictly needed. Sometimes this makes it easier to expand an example later, or sometimes the example illustrates some good coding practice that will become more relevant when you expand on the example in your own projects.

On the other hand, these examples often use literal numbers where named constants might be better practice, to make the code snippets in the book more self-contained and concise.

Variable types

The example sketches in this chapter use some variable types that you might not be familiar with. You have probably seen normal numeric types, such as char, int, or long. A problem with these types is that their size (and thus the range of values they can store) depends on the compiler and processor architecture used.

To prevent surprises, this code uses more explicit types: uint8_t, uint16_t, uint32_t, and uint64_t for unsigned (non-negative) integers of 8-, 16-, 32-, and 64-bits long and int8_t, int16_t, int32_t, and int64_t for the signed (negative or positive) versions. size_t is an unsigned integer that is guaranteed to be big enough to store the size of an object, array, and so on.

PROGMEM and F() strings

On most Arduinos any constant strings used in your code are stored in its program memory, but copied to the dynamic memory (RAM) on startup. This is needed because the code normally expects to find strings in RAM, even when the strings are not going to be changed. Since this is typically a waste of memory, the examples in this book use the F() macro and sometimes the PROGMEM keyword to prevent copying data to RAM. This only works with code that expects the strings to live in program memory, but the F() macro allows code (such as the printing functions) to handle strings in RAM as well as program memory.

For more details about this, see http://www.gammon.com.au/progmem. In short, make sure that, whenever you are printing a string, you wrap it in F(), like:

```
Serial.println(F("This string uses no RAM!"));
```

Pointers

Some of the code in the subsequent chapters will use pointers (variables that contain the address of another variable) as well. If you are not familiar with them, they can be confusing. The examples in this book only make minimal use of them and you should be able to follow the examples without fully understanding pointers. However, if you are ever confused by code that uses the &, * or -> pointer operators, it might be good to read up on pointers. There is a good introductory tutorial here: http://www.cplusplus.com/doc/tutorial/pointers/.

Serial port setup

As noted before, all of the example sketches are written for using AltSoftSerial library, but have a bit of code that makes it easier to use a different serial port setup too. That code looks like this:

```
#include <AltSoftSerial.h>
AltSoftSerial SoftSerial;
#define DebugSerial Serial
#define XBeeSerial SoftSerial
```

First, this declares a AltSoftSerial instance named SoftSerial. Then it uses #define to define two preprocessor macros. Essentially, this tells the compiler: wherever you see DebugSerial, read that as Serial instead, and wherever you see XBeeSerial, read that as SoftSerial. In the rest of the code, the DebugSerial name is used whenever some debug message must be printed and XBeeSerial is used to talk to the XBee device. For more details on the preprocessor, see the link: http://www.cplusplus.com/doc/tutorial/preprocessor/.

This approach lets you quickly modify an example to run with a different serial configuration, without having to change code throughout the entire sketch. For example, to permit an example to work on an Arduino Leonardo, using the virtual USB serial port `Serial` for debug output and the hardware serial port `Serial1` for talking to the XBee module, the preceding lines would be replaced with:

```
#define DebugSerial Serial
#define XBeeSerial Serial1
```

In the code bundle, `Connect_Leonardo.ino` is provided with those changes applied, to give you a full working example for a Leonardo or Fio v3 setup. All subsequent examples will only be provided for the AltSoftSerial setup, but you should be able to adapt those to your needs now.

The xbee-arduino library

By the end of the previous chapter, you have been using **API frames** to send data to — and receive data from — your XBee modules. So far, you have used the XCTU program to construct these packets. On the Arduino, there is a library called **xbee-arduino** that can handle building, sending, receiving, and parsing API frames for you.

The central part of this library is a number of request and response objects that represent API frames sent to or received from the XBee module respectively. There is one of these objects for every API frame supported by the library. Each object allows setting or getting the data contained in it through a number of methods calls. The request and response objects used in these examples will be introduced when they are first used, but refer to the xbee-arduino documentation for a full list.

The xbee-arduino library can be installed through the library manager, or found on github:

```
https://github.com/andrewrapp/xbee-arduino
```

The documentation for this library can be found in the same place. Go ahead and install the library now, so you can start writing the first sketch.

Creating the sketch

With the preparation done, it is time to write the sketch. It is fairly simple, containing a bit of global initialization and just a `setup()` and `loop()` function.

The global initialization starts with the includes:

```
#include <XBee.h>
#include <Printers.h>
#include <AltSoftSerial.h>
```

This sets up the `include` headers for the xbee-arduino and AltSoftSerial libraries. It additionally includes `Printers.h`, a header file also provided by the xbee-arduino library, that makes a number of extra (debug) printing functions available.

It continues by setting up a global `xbee` object, which will be used to access all xbee-arduino functionality:

```
XBeeWithCallbacks xbee;
```

Finally, the serial objects to be used in this sketch are defined, as you saw previously:

```
AltSoftSerial SoftSerial;
#define DebugSerial Serial
#define XBeeSerial SoftSerial
```

All of this global initialization code will be needed for every sketch that talks to an XBee module, so make sure to include this in all your XBee sketches. These lines will not be repeated for every example in the rest of this book, though the full example files in the code bundle will of course contain them.

The `setup()` function begins by starting up the debug and XBee serial connections:

```
void setup() {
  // Setup debug serial output
  DebugSerial.begin(115200);
  DebugSerial.println(F("Starting..."));
  // Setup XBee serial communication
  XBeeSerial.begin(9600);
  xbee.begin(XBeeSerial);
  delay(1);
```

`DebugSerial` is prepared just as you would normally prepare `Serial` for debug output (remember that the compiler actually replaces `DebugSerial` with `Serial` with the default defines shown previously).

Then, `XBeeSerial` object is similarly initialized and the `xbee` object is connected to it. Note that there is also a small delay present; this makes sure that the serial line has a bit of time to settle into its idle state before the first bytes are sent. It depends on the shield used if this delay is really required, but it should not do any harm in any case.

Again, this part of the setup function will be present in all future sketches that use the XBee module (unlike the next part, which will be changed in future sketches).

The `setup()` function continues as follows:

```
    // Let all responses be printed
    xbee.onResponse(printResponseCb, (uintptr_t)(Print*)&DebugSerial);
    // Send a "VR" command to retrieve firmware version
    AtCommandRequest req((uint8_t*)"VR");
    xbee.send(req);
}
```

The first line sets up the XBee library to print all responses received, by setting up a callback function (what a callback is and how they work will be detailed later).

The last lines send an **AT command** API frame containing the VR command. Since no value is passed along with the command, this requests the current VR value, which is the XBee firmware version.

The final part of this sketch is the `loop()` function:

```
void loop() {
    // Check the serial port to see if there is a new packet available
    xbee.loop();
}
```

Here, the `xbee.loop()` function is called, which allows the xbee-arduino library to regularly poll the serial port for new data and process it. Again, this line should be present in all future XBee sketches, even if not explicitly shown in the book.

With this sketch, you should be able to check whether the connection to the XBee module is working correctly. If so, the serial monitor should show something like:

```
Starting...
AtCommandResponse:
    Command: VR
    Status: 0x00
    Value: 23 A7
```

This shows the response to the VR command, which was successful (status is zero) and returns the 23A7 firmware version.

Even though this is of course just a simple example, you might find that this particular sketch can be useful later, too, since it prints a detailed dump of any API frames received, including the raw bytes in any received radio packet.

Sending and receiving data

Now that the XBee connectivity is covered, it is time to actually exchange some data wirelessly between your Arduinos. In the previous chapter, you have already seen the API frames involved in transmitting and receiving data through the network. In this section, you will let the Arduino use those API frames to exchange a simple `Hello, world!` message again. For this, the sender will use the XBee module that you configured as a router, while the receiver will use the coordinator XBee.

Sending data

Remember that sending a message involves a few different API frames:

- The sender sends a **ZigBee Transmit Request** API frame to the XBee module, containing the destination address and the message.

- The receiver receives a **ZigBee Receive Packet** API frame, containing the sender address and the message. The receiver XBee will also (automatically) send an acknowledgement (ack) back to the sender so the sender knows the transmission was successful.

- The sender receives a **ZigBee Transmit Status** API frame from the XBee module. This indicates whether the transmission was successful (when the acknowledgement is received) or unsuccessful (when no acknowledgement was received after a timeout, or no route could be found, and so on).

This process is illustrated in the sequence diagram next. Each of the vertical lines indicates a component in the system, and arrows indicate messages exchanged between them. To read it, just start at the top and move down to see what messages are exchanged in what order:

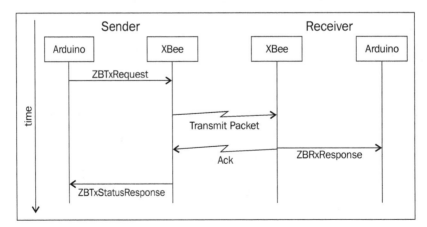

Note that **Transmit packet** is simplified in this diagram. In reality, there might be more messages exchanged between the two XBee modules to handle address resolution, routing, and retries. Since all of those are handled by the XBee module transparently, you can just think of the module sending a single radio message.

The ZBTxRequest objects

In the xbee-arduino library, **ZBTxRequest** objects are used to represent **ZigBee Transmit Request** API frames. These packets are used to send data to other XBee modules wirelessly. A transmit request is always answered with a **ZigBee Transmit Status** response.

A transmit request has a number of fields, of which these will be used:

- A 64-bit destination address. This address is mandatory and indicates where to send the packet. The special value 0x0000000000000000 can be used to address the coordinator (regardless of its actual address) and 0x000000000000FFFF can be used to send a message to all other nodes. This field is set using setAddress64().

- A 16-bit destination address. This address is optional and can be used if your sketch already knows the 16-bit address of the destination, saving the overhead of doing an address lookup on the XBee module (but even then the 64-bit address has to be also specified). In general, this can be left at the default value of 0xFFFE to indicate the address is unknown. This field is set using setAddress16().

- The payload. This is the actual content for the message, specified in two parts: a pointer to the data to be sent, as well as the length of the data (number of bytes). Both can be set using setPayload().

 The name of this object is slightly misleading, since it is also used for the DigiMesh and 868 modules, not just the XBee ZB family (but then the 16-bit address is unused).

If you are using the XBee 802.15.4 modules, you will need to use a Tx16Request or Tx64Request object instead.

Creating the sketch

The sketch for sending data is called `HelloSend.ino` in the code bundle. Using what you learned previously, you should be able to write a `sendPacket()` function that transmits your `Hello, world!` packet:

```
void sendPacket() {
    // Prepare the Zigbee Transmit Request API packet
    ZBTxRequest txRequest;
    txRequest.setAddress64(0x0000000000000000);
    uint8_t payload[] = {'H', 'e', 'l', 'l', 'o',
                        ',', ' ', 'w', 'o', 'r', 'l', 'd', '!'};
    txRequest.setPayload(payload, sizeof(payload));
    // And send it
    uint8_t status = xbee.sendAndWait(txRequest, 5000);
    if (status == 0) {
      DebugSerial.println(F("Succesfully sent packet"));
    } else {
      DebugSerial.print(F("Failed to send packet. Status: 0x"));
      DebugSerial.println(status, HEX);
    }
}
```

This function firsts creates the `ZBTxRequest` to send. The destination address is set to the all-zeroes value, sending the message to the coordinator. If you want to send the message to another XBee node that is not the coordinator, put its extended address here. This also applies when not using XBee ZB modules, since the other families do not have a special coordinator (address) like XBee ZB has.

The payload of a packet is the actual data to transmit. As far as the XBee module is concerned, this is just a bunch of bytes that are transmitted as-is. In this example, an array of bytes (`uint8_t`) is used, but a pointer to any piece of memory can be used as the payload.

Actually transmitting a packet uses `xbee.sendAndWait()`. In the previous example, you saw `xbee.send()`, which sends the request to the XBee module and immediately returns. `xbee.sendAndWait()` additionally waits for a reply from the XBee module. Not all requests will receive a reply, but transmit requests such as these do.

The status value is automatically extracted from the reply API frame and returned. The status is always `0` for success; non-zero values indicate failure (the exact values used depend on the reply API frame type and can be found in the XBee product manual).

The second argument to `xbee.sendAndWait()` is how long to wait for the response. With the default XBee configuration, a packet transmission will be tried up to three times, lasting up to 4800ms in total, so you should receive a status within this time. The preceding code stops waiting if no status reply was received within 5000ms, just in case the reply is somehow never sent.

Now you defined a `sendPacket()` function, but of course it must be called before it will actually do anything. To send a message every 10 seconds, you might think that something like the following would solve this:

```
void loop() {
  xbee.loop();
  sendPacket();
  delay(10000);
}
```

However, this would mean that `xbee.loop()` is only called once every 10 seconds. If, during those 10 seconds, any serial data is received from the XBee module, the xbee-arduino library cannot handle that data properly, potentially causing unexpected behavior. The problem here is that `xbee.loop()` needs to be regularly called to poll for new data but `delay()` blocks execution, preventing the polling from happening.

To fix this, you should not use `delay()`, but instead regularly check if it is time to send a new packet again. If you have been looking through the example sketches on the Arduino site, you might recognize this is the same approach as used by the `Blink without delay` example:

```
unsigned long last_tx_time = 0;
void loop() {
  // Check the serial port to see if there is a new packet available
  xbee.loop();
  // Send a packet every 10 seconds
  if (millis() - last_tx_time > 10000) {
    last_tx_time = millis();
    sendPacket();
  }
}
```

This keeps a global variable `last_tx_time` to track the time of the last message sent. Using this value, the time since the last transmission is continuously calculated. As soon as this is more than 10 seconds, a packet is transmitted.

After about 50 days, the value returned by millis() will overflow (restart at 0). The preceding code will handle this without a problem. See this article for more details on how this works exactly: http://www.gammon.com.au/millis.

To make sure that a packet is transmitted directly at startup (instead of waiting for 10 seconds), add a call to sendPacket() at the end of the setup() function too.

Blocking and polling

When waiting for and responding to external events (such as a changing the I/O pin, or an incoming API frame on the serial port), the most common method used is **polling**. Polling means to regularly check a pin, read the serial buffer, query an external device, and so on to see if anything changed or if there is data ready to be processed.

When polling, it is usually important to keep polling quickly. If there is too much time between two subsequent polls, events will be handled too late and in some cases events might be missed altogether.

When a piece of code runs for a longer time (because it contains a delay(), waits for some external condition, or simply has a lot of calculations to make), it is said to be **blocking**. As you can probably imagine, blocking can lead to problems when combined with polling.

For this reason, it is often good to avoid blocking code such as delay() and instead use small pieces of code that just check a condition and/or do a small piece of work, and then quickly return again. In the preceding example, the code is essentially polling the millis() value instead of blocking using the delay() function.

Another approach is to use interrupts, where some events cause a special function to be called that can interrupt even blocking code. For more information about interrupts, see this excellent article: http://www.gammon.com.au/interrupts.

Running the sketch

Now upload this sketch to your Arduino. If you open up the serial console (select 115200 baud, as specified in the sketch), it should output a message every 10 seconds:

```
Starting...
Failed to send packet. Status: 0x21
Failed to send packet. Status: 0x21
Succesfully sent packet
Succesfully sent packet
```

In the preceding output, the coordinator was initially off. This resulted in a transmit status of 0x21, which means `Network ACK Failure` (for the full list of status codes, see the product manual, under *API operation -> API frames -> ZigBee Transmit Status*). After two packets, the coordinator was switched on and packets were successfully sent.

To further verify that the message is actually being received too, you can upload the `Connect.ino` sketch to another Arduino, with the coordinator XBee module connected, and then the serial monitor should look something like this:

```
Starting...
AtCommandResponse:
   Command: VR
   Status: 0x00
   Value: 21 41
ZBRxResponse:
   From: 0x0013A20040E2C832
   From: 0x4FC6
   Receive options: 0x01
   Payload:
      48 65 6C 6C 6F 2C 20 77
      6F 72 6C 64 21
```

This shows the received packet coming in as a `ZBRxResponse`. The payload is shown in hexadecimal, but try looking up the values in an ASCII table (such as on `http://www.asciitable.com/`) and you will see this is really the `Hello, world!` message sent by the other module.

Callbacks

To simplify the processing of received data, the xbee-arduino library supports **callback functions** (or just callbacks). Generally speaking a callback function is a function that is first passed to some function to **register** the callback. Later, the callback function will be called when some specific event happens.

You have already seen a callback function being used in the `Connect.ino` example:

```
xbee.onResponse(printResponseCb, (uintptr_t)(Print*)&DebugSerial);
```

This line consists of three parts:

1. The **registration method**, `xbee.onResponse()`. This is the method that is actually being called in this line of code. This method registers a new callback function to be called when any API frame is received from the XBee module.

2. The **callback function**, printResponseCb. This is the (name of the) actual function that will be called later. In this case, it is a callback function pre-defined by the xbee-arduino library, but you will later see how to define your own callback functions as well.

3. **userdata**, (uintptr_t)(Print*)&DebugSerial. This last (optional) argument to the registration method can be used to pass a bit of extra data to the callback function.

   In this case, userdata is used to tell the printResponseCb callback function that it should print to the DebugSerial serial port.

   If this does not make any sense to you now, that is okay: you will not need to use the userdata argument for any of your own callbacks for the examples in this book. Just remember that, whenever you use any of the printSomethingCb callback functions from the xbee-arduino library, you should just make sure to pass in the argument as shown previously (if you forget, nothing will be printed).

   Explaining the exact meaning of the argument as shown is beyond the scope of this book, but check out the xbee-arduino documentation for more details.

Instead of using callbacks, it is also possible to query the xbee-arduino library for any received packets directly. Since using callbacks gives cleaner code, only the callback style will be used in this book, but be aware that a lot of the examples you might find will use the other (older) style.

Callback types

A number of callback registration functions is available, each registering callbacks to be called for different events:

- onResponse() registers a callback that will be called for every response (API frame) received from the XBee module. This happens in addition to calling any of the response-specific callbacks or the onOtherResponse() callback described next.

- For every kind of response, a response-specific callback can be registered. For example, using onTxStatusResponse() you can register a callback to be called when a TxStatusResponse (**Transmit Status** API frame) is received, while the onZBRxResponse() callback is called for a ZBRxResponse (**ZigBee Receive Packet** API frame). Any of these callbacks used in the examples in this book will be introduced when needed; refer to the xbee-arduino documentation for a full list.

- `onOtherResponse()` will be called for every response received for which no response-specific callback was registered. This can be used to log unexpected responses, for example.

- `onPacketError()` will be called when an error occurs while reading an API frame (such as a checksum error).

Callback limitations

In order to make sure that these callbacks are actually called, `xbee.loop()` should be regularly called, so it can check for new incoming data. When you use `xbee.sendAndWait()`, as you saw previously, this happens automatically and the callbacks will be called for any API frames received while waiting for the reply.

Additionally, not everything is allowed inside a callback. In general, a callback should never take very long or block, and should certainly not call `xbee.sendAndWait()`, since that might call another callback, messing up the library's internal state.

Receiving data

Now that you have one Arduino sending out radio packets, it is time to improve the reception code. You have used the `Connect.ino` sketch to print the received packets, but the code that actually handles the received messages is hidden away in the `printResponseCb` callback inside the xbee-arduino (and it only prints the data in hexadecimal format).

To allow reading the `Hello, world!` bytes you have transmitted, but more importantly to prepare for receiving and processing meaningful data later, you will need more control over the reception of packets.

As you have seen in the previous chapter, whenever a packet is received, the XBee module sends a **ZigBee Receive Packet** API frame, so you will need to register a callback to process those.

The ZBRxResponse objects

In the xbee-arduino library, `ZBRxResponse` objects are used to represent **ZigBee Receive Packet** API frames. These API frames are used whenever the XBee module receives a radio message. To process `ZBRxResponse` objects, you can register a callback using `onZBRxResponse()`.

The ZBRxResponse objects expose a number of methods to allow access to their data:

- getRemoteAddress16(): Returns the 16-bit short address of the sender.
- getRemoteAddress64(): Returns the 64-bit extended address of the sender.
- getDataLength(): Returns the number of data bytes in the message.
- getData(): Provides access to the data bytes in the message. Calling getData() with no arguments returns a pointer to the entire message. This pointer can then be passed to another piece of code, or indexed with the [] operator just like an array. Alternatively, passing a single integer to getData() lets it return just the single byte at the given position.
- The getOption(): Returns the receive options for this message. See the XBee product manual for the meaning of this value.

Again, this same object is used for the DigiMesh and 868 modules, while the XBee 802.15.4 modules use an Rx16Response or Rx64Response instead.

Creating the sketch

The sketch for receiving data is called HelloReceive.ino in the code bundle. It again contains all the standard code you have seen in Connect.ino, but has a different callback registration in setup():

```
xbee.onZBRxResponse(processRxPacket);
```

The callback function registered is processRxPacket(), which simply prints any received messages to the serial console:

```
void processRxPacket(ZBRxResponse& rx, uintptr_t) {
  DebugSerial.print(F("Received packet from "));
  printHex(DebugSerial, rx.getRemoteAddress64());
  DebugSerial.println();
  DebugSerial.print(F("Payload: "));
  DebugSerial.write(rx.getData(), rx.getDataLength());
  DebugSerial.println();
}
```

Note that this function has a second uintptr_t parameter without a name; this is the userdata parameter mentioned earlier. Even if you do not use it, it must be declared. By specifying the type, but no name for the parameter, the compiler knows that you are not planning to use it and will not emit an unused parameter warning.

If you upload this sketch, it should start dumping any received radio packets to the serial console. If everything worked out correctly, this is what you should be seeing now:

```
Starting...
Received packet from 0013A20040E2C832
Payload: Hello, world!
Received packet from 0013A20040E2C832
Payload: Hello, world!
```

Collecting sensor data

Even though sending over `Hello, world!` is already pretty cool, it is not very useful yet. Instead of sending a fixed message, you will want to send some variable data, such as coming from a sensor.

In this section, you will connect a combined temperature and humidity sensor to the sending Arduino and have it send over its readings to the coordinator. The coordinator will then receive these readings and, for now, display them on the Serial console (later, you will see how to put the readings in pretty graphs, too!).

There are a lot of different sensors available that can be used with an Arduino. There is not a single unified way to wire up and talk to all sensors, but Arduino-specific instructions and tutorials can be found for a lot of hardware. A good source of inspiration are Arduino-oriented online shops such as http://www.sparkfun.com or http://www.adafruit.com. They stock all kinds of sensors and typically provide appropriate Arduino libraries and/or examples as well.

The sensor used in this example is the DHT22, but there are a lot of other sensors that you can use instead. The library used (see later) supports a number of similar sensors with minimal changes to the code, and of course completely different sensors (even sensing different things) can be used as well, though that requires more modifications to the code not described in this book.

Next, you will start with just reading the sensor itself, to verify that it is connected correctly. Then you will modify the `HelloSend.ino` example shown previously to send the sensor values and finally modify the `HelloReceive.ino` example to receive and print the sensor values.

Reading a DHT22 sensor

The first step is to wire up the DHT sensor. When looking at the sensor from the front (the half-open side), the pins are (from left to right): vcc, data, unused, and ground. Vcc and ground connect to the 5V and GND pins on the Arduino. The data pin can connect to any I/O pin on the Arduino; this example uses pin 4. The data pin needs a pullup resistor too. Usually the internal pullup (inside the Arduino, it is automatically enabled by the DHT library) should work, but it is a bit weak. So if you run into problems reading the sensor, adding an external resistor (around 10k — the exact value does not really matter) might help.

For more info on these sensors, see this page by Adafruit: `https://learn.adafruit.com/dht`.

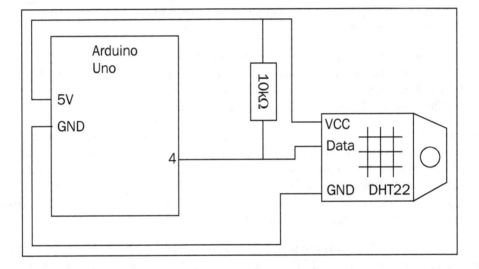

Note that you can just leave the XBee shield connected and plug the wires into the shield. Using a small breadboard is probably a good idea, too.

To actually read the sensor values from your code, you will need a library that knows about these DHT sensors. There are a few of them out there, but this example uses the **DHT Sensor Library** by Adafruit:

```
https://github.com/adafruit/DHT-sensor-library
```

This library can be easily installed through the Arduino library manager, or manually downloaded from github.

Next up is writing some code. Create a Dht.ino sketch and start by adding some initialization code:

```
#include <DHT.h>
// Sensor type is DHT22, connected to pin D4.
DHT dht(4, DHT22);
#define DebugSerial Serial
void setup() {
  // Setup debug serial output
  DebugSerial.begin(115200);
  DebugSerial.println(F("Starting..."));
  // Setup DHT sensor
  dht.begin();
}
```

This includes the DHT library and creates a global DHT object. It is configured to use digital pin 4 as the data pin and to talk to a DHT22 sensor (change this value if you are using a different sensor).

In the setup() function, dht.begin() is called to initialize the library. After that, you should be ready to read some values in the loop() function:

```
void loop() {
  DebugSerial.print(F("Temperature: "));
  DebugSerial.println(dht.readTemperature());
  DebugSerial.print(F("Humidity: "));
  DebugSerial.println(dht.readHumidity());
  delay(10000);
}
```

If you upload this sketch and wired up the sensor correctly, you should get something like this on the serial console:

```
Starting...
Temperature: 21.40
Humidity: 49.50
```

Great! You can now read the temperature and humidity! Now it is time to start sending these readings into the air.

The sketches in this book use floating point numbers to represent sensor readings, since that is the easiest way to handle fractional numbers. The downside is that floating point handling is much less efficient than integer handling on most Arduinos, taking up more program storage space and computation time. If this is a problem, consider alternative approaches, such as using an integer value to represent a sensor reading in tenths of a degree.

Handling packets using binary.h

In the code bundle for this book a file called binary.h is supplied that contains some code to help with working with binary packet data. This file defines two class types—Buffer and AllocBuffer—that can be used to construct or deconstruct a binary packet.

Next, some examples are shown of how to use this file, and the rest of the examples in this book also make use of it. The implementation of these class types is not covered in this book, but you are encouraged to look inside binary.h to learn more about how that code works.

Memory representation and byte order

The code used in binary.h copies variables directly from memory into the packet. This means that the way these variables are stored in memory defines how they end up in the packet. However, different processor architectures and compilers use different ways of representing values in memory (the most notable difference is big endian byte order versus little endian byte order).

Most, if not all, Arduino boards use the little endian byte order and the same memory layout, making them interoperable. If you ever include different kinds of hardware in the network, be aware that some additional work might be needed (such as converting to the right byte order).

Building and sending a packet

To create a packet, you will be using some code from binary.h, which will be explained first. Then, you will need to send a **ZigBee Transmit Request** API frame and send it to the module, which will then take care of sending your message along to the destination.

Constructing a packet using binary.h

To construct a packet, you typically use `AllocBuffer`, which will allocate a region of memory where your packet can be constructed. Declaring such a buffer typically looks like this:

```
AllocBuffer<9> packet;
```

This means creating a new, empty `AllocBuffer` that will be nine bytes in size. Then, to add data to the buffer, you can use the `append()` method:

```
packet.append<uint8_t>(0x10);
packet.append<float>(1.0);
packet.append<float>(999.9);
```

Each call to `append()` adds the given data to the buffer and internally tracks how much data has been added so far and where the next piece of data will be added. Note that the type of the data to store is explicitly specified — in this case, an unsigned 8-bit integer and two floating point values. Note that it is up to you to make sure that your buffer has enough space to hold all the values (nine bytes in this case, a 1-byte `uint8_t` value plus two 4-byte `float`s). If you append more data than there is room, the extra data will be silently dropped.

When you have added all needed content to the packet, you can refer to the constructed packet using the `head` attribute and the `len()` method:

```
uint8_t *ptr = packet.head;
size_t length = packet.len();
```

Typically, you would pass these two values to the code that will actually send the packet.

> The pointy brackets used in this code indicate that a **C++ template** is being used, for example to allow a single function to work for various argument types. How these work exactly is beyond the scope of this book, but see this introduction if you are curious: `http://www.codeguru.com/cpp/com-tech/atl/tutorials/article.php/c3617/An-Introduction-to-C-Templates.htm`.

Creating the sketch

The sketch for sending DHT sensor data is called `DhtSend.ino` in the code bundle. It is mostly identical to the `HelloSend.ino` sketch with some minor changes.

The initialization of this sketch combines the `HelloSend.ino` and `Dht.ino` sketch:

```
#include <XBee.h>
#include <Printers.h
#include <AltSoftSerial.h>
#include <DHT.h>
#include "binary.h"
XBeeWithCallbacks xbee;
AltSoftSerial SoftSerial;
#define DebugSerial Serial
#define XBeeSerial SoftSerial
// Sensor type is DHT22, connected to pin D4.
DHT dht(4, DHT22);
```

The only addition here is including `binary.h`, which defines some helpers for building a binary packet. This file is supplied in the code bundle accompanying this book (along with the `HelloSend.ino` sketch), so make sure you copy it into your sketch directory.

The `setup()` function is the one from `HelloSend.ino` and `Dht.ino` combined, except that some additional callback functions are registered:

```
xbee.onPacketError(printErrorCb, (uintptr_t)(Print*)&DebugSerial);
xbee.onResponse(printErrorCb, (uintptr_t)(Print*)&DebugSerial);
```

These two lines make sure that, when any error happens when receiving an API frame or when any API frame is received that indicates an error, a debug message is printed. This ensures that, while testing, any errors will not go undetected and allows for simplifying the rest of the code a bit. These two callbacks will be used in all future examples as well.

The sending of a packet is now somewhat more involved. Where the `HelloSend.ino` example just had a single array of bytes to be sent, you now have two different `float` values to be sent. Some kind of **packet structure** has to be designed to hold these values. Essentially, the packet structure is an agreement between the sender and receiver about the meaning of all bits and bytes in the payload.

For now, you could just stuff the two float values into the payload and send that over. However, if you later want to add other devices that send other types of data too, receivers will have no way to tell what kind of packet they have received and how to interpret the data inside.

To solve this, all packets sent will start with a single type byte. For the packets sent by this sketch, this type byte has a value of 1 (this intentionally does not use the value 0; you will see why next). Different packets that will be defined later (such as in *Chapter 4, Controlling the World*) will use different values for this type byte.

The rest of the packet payload is fairly simple, it is just the temperature and humidity floating point values concatenated together. The resulting packet structure can be summarized as:

0	1	2	3	4	5	6	7	8
type (1)	temperature (float)				humidity (float)			

The sending of the packet is similar to before: creating and sending a ZBTxRequest object. The generation of the payload array is different now:

```
void sendPacket() {
    // Prepare the Zigbee Transmit Request API packet
    ZBTxRequest txRequest;
    txRequest.setAddress64(0x0000000000000000);
    // Allocate 9 payload bytes: 1 type byte plus two floats of 4
    // bytes each
    AllocBuffer<9> packet;
    // Packet type, temperature, humidity
    packet.append<uint8_t>(1);
    packet.append<float>(dht.readTemperature());
    packet.append<float>(dht.readHumidity());
    txRequest.setPayload(packet.head, packet.len());
    // And send it
    xbee.send(txRequest);
}
```

This uses the `AllocBuffer` type, as described earlier, to allocate the nine bytes needed for the payload. The buffer is then filled using the type byte and two floating point values.

After adding the data to the buffer, a pointer to the buffer is stored in `txRequest` using `setPayload()`. Note that this does not make a copy of the data; it just makes the request point to the data in the buffer. This means you need to be careful to not let the `AllocBuffer` object go out of scope before the request object is sent, since the latter will then point to invalid memory.

Note that this no longer waits for the transmit status and prints a success or error message (like `HelloSend.ino` did). Instead, it relies on the `printErrorCb` registered in `setup()` to print an error when transmission fails.

The `loop()` function is identical to the one from `HelloSend.ino`, so you should now have a complete sketch that reads sensor values and sends them to the coordinator through the radio. You can use the `Connect.ino` sketch on the coordinator again to confirm that data is being sent. The ASCII values printed will look like gibberish, but you should be able to recognize the `0x01` type byte, followed by eight more bytes (four for each float).

This example sends a reading every 10 seconds, but once things are working, you should probably increase this interval. The rest of this book assumes that readings are sent once every 5 minutes.

Receiving and parsing a packet

The final piece of this puzzle is to receive the sensor data on your coordinator Arduino. You have already seen how to receive a packet using the `onZBRxResponse()` callback in the `HelloReceive.ino` example, so the only missing piece is how to parse the received packet.

Parsing a packet using binary.h

Parsing a packet refers to deconstructing the raw received bytes into separate values again. This can be done using the `Buffer` class from `binary.h`. Assuming that you have a pointer to the raw packet in a variable called `ptr` and the length of the packet in `length`, you would create the buffer like this:

```
Buffer packet(ptr, length);
```

After that, you can remove values from the buffer one by one using the `remove()` method:

```
uint8_t value1 = packet.remove<uint8_t>();
float value2 = packet.remove<float>();
float value3 = packet.remove<float>();
```

You should take care to remove things from the buffer in the same order as they were added, otherwise the deconstructed values will not match the original values sent (and the compiler cannot check this for you).

The length of the remaining data in the buffer can be retrieved using the `len()` method again. Note that this only counts remaining data, so this value decrements whenever you call the `remove()` method.

Generally, when deconstructing a packet like this, you should carefully check the length before reading from the packet, to ensure you are not reading past the end of the buffer and reading from a random piece of memory instead. When using the `Buffer` class, this risk is reduced: It will never read past the end of the valid data but, if you try to remove more data than is present, it will return only zeroes.

This means that, if a zero value can be interpreted as a valid packet, you should still check the length beforehand. However, if a zero value is invalid and will trigger other error handling, it is not needed to separately check the packet length too. The packet type byte used in the `Coordinator.ino` example applies this approach.

Creating the sketch

The sketch to receive your sensor values is called `Coordinator.ino` in the code bundle. It also needs `binary.h`, so make sure that the file is present in the sketch directory as well.

This sketch is very similar to the `HelloReceive.ino` sketch, except that it should include `binary.h` at the top and of course the `processRxPacket()` function will be different. To make it receive the sensor data, it has to handle the packet structure defined by the preceding sender:

```
void processRxPacket(ZBRxResponse& rx, uintptr_t) {
  Buffer b(rx.getData(), rx.getDataLength());
  uint8_t type = b.remove<uint8_t>();
  if (type == 1 && b.len() == 8) {
    DebugSerial.print(F("DHT packet received from "));
    printHex(DebugSerial, rx.getRemoteAddress64());
    DebugSerial.println();
    DebugSerial.print(F("Temperature: "));
    DebugSerial.println(b.remove<float>());
    DebugSerial.print(F("Humidity: "));
    DebugSerial.println(b.remove<float>());
    return;
  }
  DebugSerial.println(F("Unknown or invalid packet"));
  printResponse(rx, DebugSerial);
}
```

This starts by creating a `Buffer` object, passing in the payload data from the `ZBRxResponse` object.

The packet is then parsed, starting with the type byte. If the type byte is 1 (indicating a packet sent by the DhtSend.ino sketch) and the payload contains exactly eight more bytes (this is two float values; the type byte is no longer included in this amount), the sender address and both float values are printed.

If the type is not recognized as a valid type, or the packet is not long enough to contain two floating point values, an error message is printed (including all details about the received packet, using the printResponse() function from the xbee-arduino library).

Note that the length is not checked before removing the type byte. If the received packet somehow has an empty payload, this code would still work as expected: remove<uint8_t>() will return a type of 0 when the packet is too short. Since a type of 0 is intentionally not a valid type, an error message is shown indicating this is an invalid packet.

When receiving a valid packet, the output on the serial monitor should look something like this:

```
Starting...
DHT packet received from 0013A20040DADEE0
Temperature: 20.80
Humidity: 58.80
```

Troubleshooting

In an ideal world, everything works as expected right away. In the real world, you will run into problems every now and then. Here are some pointers to help you figure out what is wrong and how to fix it.

Communication with the XBee module is not working

Is serial communication not working at all? Check your connections: DIN to TX, DOUT to RX.

A good way to test this is to do a loopback test:

1. Remove the XBee module from your board and use a jumper wire to connect DOUT (pin 2) and DIN (pin 3) on the board.

2. Upload the SerialDump.ino sketch to the Arduino and open up the serial port using the serial monitor in the Arduino IDE (or another serial console program).

3. If everything is wired correctly, any data you send should be received and displayed again.

If you are using a separate USB-to-serial converter, try swapping the DIN and DOUT wires; some of these converters have their labels inverted.

Make sure you connect GND between the XBee and USB-to-serial converter too, to establish a common ground reference.

Is the XBee module getting 3.3V power from somewhere? Shields and USB adapters usually have this covered but, when using simpler breakout boards, you typically have to take care of supplying power yourself.

Do you have the right serial settings? Maybe you changed the baud rate in the past? Use the device discovery tool in XCTU to try a lot of different settings at once.

Check you did not leave the serial port open in some other program (like the Arduino serial monitor for testing, perhaps?). Opening the same port twice does not always give an error message, but does prevent data from being received properly.

Are permissions set up correctly? On Linux, you might not be allowed access to virtual serial ports. See http://playground.arduino.cc/Linux/All#Permission for more info.

Modules are not joining the network

Are your modules not joining the network? Compare settings; is the PAN ID (ID) matching, are there matching security settings (EE) and keys (KY)? Does the channel mask (CH) include the current channel?

Perhaps joining is disabled; check the NJ parameter on a coordinator or router that is near the joining device.

Check the AI (Association Indication) value to see if it reports any errors. The meaning of these values can be found in the product manual for your XBee module.

Modules cannot talk to each other

A module is joined to the network successfully, but it cannot send or receive any data.

Are you using the right 64-bit destination address? Are you using the right 16-bit address (or 0xfffe to indicate an unknown address?).

If your network has security enabled, do all nodes have EE=1? If a node does not have security enabled, it will join a secure network and then just ignore the security keys it gets offered. It looks like it is joined correctly, but the node will send unencrypted packets that are ignored by the rest of the network.

Perhaps the modules are joined to different networks? Check the OI value on each module; they should be equal. If you replaced the coordinator's firmware or made a configuration change that made it start a new network, some routers might have been left in the old network. To force a router to leave the old network and join the new network, change its ID configuration value, click the **Write** button, then change ID back to the correct value and click **Write** again.

Summary

In this chapter, you have put your Arduinos in control of your radio modules, and set up your very first wireless sensor network by wirelessly transmitting measured values to your coordinator node. By applying what you have learned, you can extend your network by adding more Arduinos with DHT sensors. You should even be able to add different kinds of sensors by adding a new packet type!

Now, reading your measurements in the serial console is nice, but not very convenient. In the next chapter, you will find out how to store all this collected data on your computer, or even directly in the cloud, and how to gain more insight by generating convenient graphs out of the data.

3
Storing and Visualizing Your Data

In the previous chapter, you built a sensor module that collects temperature and humidity data and sends it wirelessly to your central coordinator. In this chapter, you will explore some ways to persistently store this collected sensor data, and to visualize the data using convenient graphs.

First you will see how to connect your coordinator to the Internet and send its data to the Beebotte cloud platform. You will learn how to create a custom dashboard inside that platform that can show the collected data in a convenient graph format.

Second, you will see how you can collect and visualize the data on your own computer instead of sending it to the Internet directly.

For the first part, you will need a shield to connect your coordinator Arduino to the Internet, in addition to the hardware recommended for the coordinator in the previous chapter. This book provides suggestions for these two shields:

- The Arduino Ethernet shield (`https://www.arduino.cc/en/Main/ArduinoEthernetShield`)
- The Adafruit CC3000 WiFi shield (`https://www.adafruit.com/products/1491`)

If possible, using the Ethernet shield is recommended, since its library is smaller and it is easier to keep a reliable connection using a wired connection. Also, the CC3000 shield conflicts with the SparkFun XBee shield, requiring some modification on the latter to make them work together.

For the second part, no additional hardware is needed.

There are other Ethernet or Wi-Fi shields you can use; see the *Connecting your Arduino to the Internet* section further on for some advice in this regard.

Storing your data in the cloud

When it comes to storing your data somewhere online, there are literally dozens of online platforms that offer some kind of data storage service aimed at collecting sensor data. Each of these has different features, complexity, and cost, and you are encouraged to have a look around at what is available.

Even though a lot of platforms are available, almost none of them are really suited for a hobby sensor network such as the one presented in this book. Most platforms support the basic collection of data and offer a Web API to access the data, but there were two requirements that ruled out most of the platforms:

1. It has to be affordable for a home user with just a bit of data. Ideally, there is a free version to get started.

2. It has to support creating a dashboard that can show data and graphs, but can also show input elements that can be used to talk back to the network (this will be used in the next chapter to create an online thermostat).

When this book was written, only two platforms seemed completely suitable: Beebotte (`https://beebotte.com/`) and Adafruit IO (`https://io.adafruit.com/`). The examples in this book use Beebotte because, at the time of writing, Adafruit IO was not publicly available and Beebotte currently has some additional features; however, you are encouraged to also check out Adafruit IO as an alternative. Since both platforms use the MQTT protocol (explained next), you should be able to reuse the example code with just minimal changes for Adafruit IO.

Introducing Beebotte

Beebotte, like most of these services, can be seen as a big online database that stores any data you send to it, and allows retrieving any data you are interested in. Additionally, you can easily create dashboards that allow you to look at your data and even interact with it through various configurable widgets. By the end of this chapter, you might have a dashboard that looks like this:

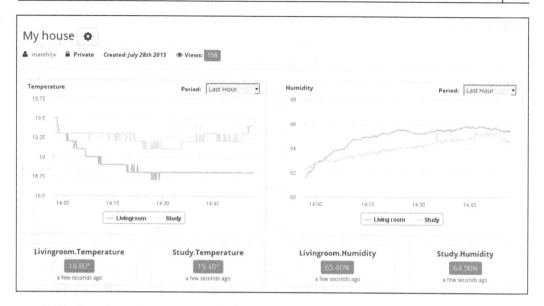

Before showing how to talk to Beebotte from your Arduino, some important concepts in the Beebotte system will be introduced: channels, resources, security tokens, and access protocols.

The examples in this book serve to get started with Beebotte, but will certainly not cover all of its features and details. Be sure to check out the extensive documentation on the Beebotte site at https://beebotte.com/overview.

Channels and resources

All data collected by Beebotte is organized into **resources**, each representing a single series of data. All data stored in a resource signifies the same thing, such as temperature in your living room or the on/off status of the air conditioner, but at different points in time. This kind of data is also often referred to as **time-series data**.

To keep your data organized, Beebotte supports the concept of channels. Essentially, a channel is just a group of resources that somehow belong together. Typically, a channel represents a single device or data source, but you are free to group your resources in any way you see fit.

In this example, every sensor module in the network will get its own channel, each containing a resource to store temperature data and a resource to store humidity data.

Security

To be able to access the data stored in your Beebotte account, or publish new data, every connection needs to be authenticated. This happens using a secret token or key (similar to a password) that you configure in your Arduino code and that proves to the Beebotte server that you are allowed to access the data.

There are two kinds of secrets currently supported by Beebotte:

- **Your account secret key**: This is a single key for your account that allows access to all resources and all channels in your account. It additionally allows creating and deleting channels and resources.

- **Channel tokens**: Each channel has an associated channel token that allows reading and writing data from that channel only. Additionally, the channel token can be regenerated if the token is ever compromised.

This example uses the account secret key to authenticate the connection. It would be better to use a more limited channel token (to limit the consequences if the token is leaked) but, since in this example the coordinator forwards data for multiple sensor nodes (each of which has their own channel), a channel token does not provide enough access.

As an alternative, you could consider using a single channel containing all resources (named, for example, `Livingroom_Temperature` to still allow grouping) so you can use the slightly more secure channel tokens. In the future, Beebotte might also support more flexible limited tokens that support writing to more than one channel.

The examples in this book use an unencrypted connection, so make sure at least your Wi-Fi connection is encrypted (using WPA or WPA2). If you are working with particularly sensitive information, be sure to consider using SSL/TLS for the connection.

Due to limited microcontroller speeds, running SSL/TLS directly on the microcontroller does not seem feasible, so this would need external cryptographic hardware, or support on the Wi-Fi/Ethernet shield used. At the time of writing, there does not seem to be any shield that directly supports this, but it seems that at least the ESP2866-based shields and the Arduino Yún could be made to support this; the upcoming Arduino Wi-Fi shield 101 might support it as well (but this is beyond the scope of this book).

Access protocols

To store new data and access existing data over the Internet, a few different access methods are supported by Beebotte:

- **HTTP/REST**: The **HyperText Transfer Protocol (HTTP)** is the protocol that powers the Web. Originally, it was used to let a Web browser request a Web page from a server, but now HTTP is also commonly used to let all kinds of devices send and request arbitrary data (instead of webpages) to and from servers as well. In this case, the server is commonly said to export an HTTP or REST (Relational State Transfer) API.

  HTTP APIs are convenient, since HTTP is a very widespread protocol and HTTP libraries are available for most programming languages and environments.

- **WebSockets**: A downside of HTTP is that it is not very convenient for sending events from the server to the client. A server can only send data to the client after the client sends a request, which means the client must poll for new events continuously.

  To overcome this, the WebSockets standard was created. WebSockets is a protocol on top of HTTP that keeps a connection (socket) open indefinitely, ready for the server to send new data whenever it wants to, instead of having to wait for the client to request new data.

- **MQTT**: The **Message Queuing Telemetry Transport** protocol (MQTT) is a so-called publish/subscribe (pubsub) protocol. The idea is that multiple devices can connect to a central server and each can **publish** messages to a given topic, but can also **subscribe** to any number of topics. Whenever a message is published to a topic, it is automatically forwarded to all devices that have subscribed to that same topic.

  MQTT, like WebSockets, keeps a connection open continuously, so both the client and the server can send data at any time, making this protocol especially suitable for realtime data and events. MQTT cannot be used to access historical data, though.

A lot of alternative platforms only support the HTTP access protocol, which works fine for pushing and accessing data and would be suitable for the examples in this chapter. However, it is less suitable for also controlling your network from the Internet, as used in the next chapter. To prepare for that, the examples in this chapter will already use the MQTT protocol, which supports both use cases efficiently.

Sending your data to Beebotte

Now that you have learned about some important Beebotte concepts, you are ready to send your collected sensor data to Beebotte. First, you will prepare Beebotte and your Arduino for connecting to each other. Then, you will write a sketch for your coordinator to send the data. Finally, you will see how to access and visualize the stored data.

Preparing Beebotte

Before you can start sending data to Beebotte, you will have to prepare the proper channels and resources to store the data. This example uses two channels: Livingroom and Study, referring to the rooms in which sensors have been placed. You should of course use names that reflect your setup and adapt things if you have more or fewer sensors.

The first step is of course to register an account on www.beebotte.com. Once you have done this, you can access your **Control Panel**, which will initially show you an empty list of channels:

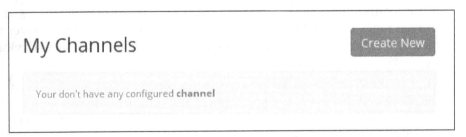

You can create a new channel by clicking the **Create New** button. In the resulting window, you can fill in a name and description for the channel and define resources. This should look something like this:

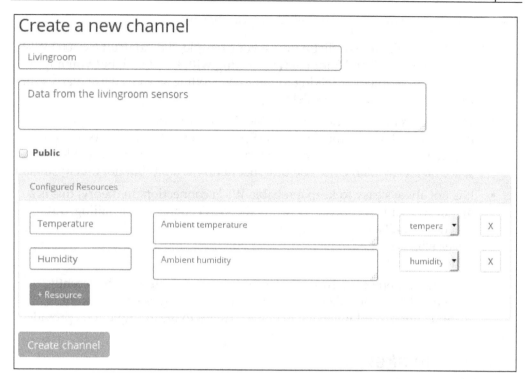

After creating a channel for every sensor node you have built, you have prepared Beebotte to receive all the sensor data. The next step is to modify the coordinator sketch to actually send the data.

Connecting your Arduino to the Internet

In order to let your coordinator send its data to Beebotte, it must be connected to the Internet somehow. There are plenty of shields out there that add wired Ethernet or wireless Wi-Fi connectivity to your Arduino. Wi-Fi shields are typically a lot more expensive than the Ethernet ones, but the recently introduced ESP2866 cheap Wi-Fi chipset will likely change that (but, at the time of writing, no ready-to-use Arduino shield was available).

Since the code for connecting your Arduino to the Internet will be significantly different for each shield, this part of the code will not be discussed in this book. Instead, we will focus on the code that connects to the MQTT server and publishes the data, assuming that the Internet connection is already set up. In the code bundle for this book, two complete examples are available for use with the Arduino Ethernet shield and the Adafruit CC3000 shield. These examples should work as a starting point for using other hardware as well.

Some things to keep in mind when selecting your hardware:

- Check carefully for conflicting pins. For example, the Adafruit CC3000 Wi-Fi shield uses pins 2 and 3 for communicating with the shield, but you might be using those pins for talking to the XBee module as well (particularly when using the SparkFun XBee shield).

- The libraries for the various Wi-Fi shields take up a lot of code space on the Arduino. For example, using the SparkFun or Adafruit CC3000 library together with the Adafruit MQTT library fills up most of the available code space on an Arduino Uno. The Ethernet library is a bit (but not much) smaller.

- It is not always easy to keep a reliable Wi-Fi connection. In theory, this is a matter of simply reconnecting when the Wi-Fi connection is failing, but in practice it can be tricky to implement this completely reliably. Again, this is easier with wired Ethernet, since that does not have the same disconnection issues as Wi-Fi.

- If you use different hardware than recommended (including the recently announced Arduino Ethernet shield 2), you will likely need a different Arduino library and will have to make changes to the example code provided.

Writing the sketch

To send the data to Beebotte, the `Coordinator.ino` sketch from the previous chapter needs to be modified. As noted before, only the MQTT code will be shown here, but, of course, the code to establish the Internet connection is included in the full examples in the code bundle (in the `Coordinator_Ethernet.ino` and `Coordinator_CC3000.ino` sketches).

This example uses the **Adafruit MQTT** library for the MQTT protocol, which can be installed through the Arduino library manager or can be found here: `https://github.com/adafruit/Adafruit_MQTT_Library`. Depending on the Internet shield, you might need more libraries as well (see the comments in the example sketches). Do not forget to add the appropriate includes for the libraries you are using. For the MQTT library, this is:

```
#include <Adafruit_MQTT.h>
#include <Adafruit_MQTT_Client.h>
```

To set up the MQTT library, you first need to define some settings:

```
const char MQTT_SERVER[]   PROGMEM    = "mqtt.beebotte.com";
const char MQTT_CLIENTID[] PROGMEM    = "";
const char MQTT_USERNAME[] PROGMEM    = "your_key_here";
const char MQTT_PASSWORD[] PROGMEM    = "";
const int MQTT_PORT = 1883;
```

This defines the connection settings to use for the MQTT connection. These are appropriate for Beebotte; if you are using some other platform, check its documentation for the appropriate settings.

Note that here the username must be set to your account's secret key, for example:

```
const char MQTT_USERNAME[] PROGMEM =
    "840f626930a07c87aa315e27b22448468844edcad03fe34f551ac747533f544f";
```

The account's secret key can be found in the Beebotte control panel, under **AccountSettings** and then **Credentials**. You will not need the **API Key** listed there for MQTT.

If you are using a channel token, it must be set to the token prefixed with `token:` such as:

```
const char MQTT_USERNAME[] PROGMEM =
    "token:1438006626319_UNJoxdmKBoMFIPt7";
```

The channel token for a channel can be found in the overview page for that channel in the Beebotte control panel.

The password is unused and can be empty, just like the client identifier. The port is just the default MQTT port number: 1883.

Note that all the string variables are marked with the `PROGMEM` keyword. This tells the compiler to store the string variables in program memory, just like the `F()` macro you have seen before does (it also uses the `PROGMEM` keyword underwater). However, the `F()` macro can only be used inside functions, which is why these variables use the `PROGMEM` keyword directly. This also means that the extra checking offered by the `F()` macro is not available; thus, be careful not to mix up normal and `PROGMEM` strings here, since that will not result in any compiler error and instead things will be broken when your run the code.

Using the configuration constants defined earlier, you can now define the main `mqtt` object:

```
Adafruit_MQTT_Client mqtt(&client, MQTT_SERVER, MQTT_PORT,
    MQTT_CLIENTID, MQTT_USERNAME, MQTT_PASSWORD);
```

There are a few different flavors of this object. For example, there are flavors optimized for specific hardware and corresponding libraries, but there is also a generic flavor that works with any hardware whose library exposes a generic `Client` object (like most libraries currently do). The latter flavor, `Adafruit_MQTT_Client`, is used in this example and should usually be fine.

The &client part of this line refers to a previously created Client object (not shown here, since it depends on the Internet shield used), which is used by the MQTT library to set up the MQTT connection.

To actually connect to the MQTT server, a function called connect() will be defined. This function is called to connect once at startup, and to reconnect every time when publishing data fails.

On the CC3300 version, this function associates to the Wi-Fi access point and then sets up the connection to the MQTT server. On the Ethernet version, where the network is always available after initial initialization, the connect() function only sets up the MQTT connection. The latter version is shown next:

```
void connect() {
  client.stop(); // Ensure any old connection is closed
  uint8_t ret = mqtt.connect();
  if (ret == 0)
    DebugSerial.println(F("MQTT connected"));
  else
    DebugSerial.println(mqtt.connectErrorString(ret));
}
```

This calls mqtt.connect() to connect to the MQTT server and writes a debug message to report success or failure. Note that mqtt.connect() returns a number as an error code (with 0 meaning OK), which is translated to a human-readable error message using the mqtt.connectErrorString() function.

Now, to actually publish a single value, there is the publish() function:

```
void publish(const __FlashStringHelper *resource, float value) {
  // Use JSON to wrap the data, so Beebotte will remember the data
  // (instead of just publishing it to whoever is currently
  // listening).
  String data;
  data += "{\"data\": ";
  data += value;
  data += ", \"write\": true}";
  DebugSerial.print(F("Publishing "));
  DebugSerial.print(data);
  DebugSerial.print(F( " to "));
  DebugSerial.println(resource);
  // Publish data and try to reconnect when publishing data fails
  if (!mqtt.publish(resource, data.c_str())) {
    DebugSerial.println(F("Failed to publish, trying reconnect..."));
    connect();
```

```
    if (!mqtt.publish(resource, data.c_str()))
        DebugSerial.println(F("Still failed to publish data"));
    }
}
```

This function takes two parameters: the name of the resource to publish to, and the value to publish. Note the type of the resource parameter: the __ `FlashStringHelper*` is similar to the more common `char*` string type, but indicates that the string is stored in program memory instead of RAM. This is also the type returned by the `F()` macro you have seen before. Just like the MQTT server configuration values that used the `PROGMEM` keyword before, the MQTT library also expects the MQTT topic names to be stored in program memory.

The actual value is sent using the **JavaScript Object Notation (JSON)** format. For example, for a temperature of 20 degrees, it constructs {`data: 20.00, write: true`}. In addition to transmitting the value, this format indicates that Beebotte should store the value so it can later be retrieved. If the `write` value is false, or not present, Beebotte will only forward the value to any other devices currently subscribed to the appropriate topic, without saving it for later. This example uses some quick-and-dirty string concatenation to generate the JSON. If you want something more robust and elegant, have a look at the ArduinoJson library at `https://github.com/bblanchon/ArduinoJson`.

If publishing the data fails, it is likely that the Wi-Fi or MQTT connection has failed, so it attempts to reconnect and publish the data once more.

As before, there is a `processRxPacket()` function, which gets called when a radio packet is received through the XBee module:

```
void processRxPacket(ZBRxResponse& rx, uintptr_t) {
    Buffer b(rx.getData(), rx.getDataLength());
    uint8_t type = b.remove<uint8_t>();
    XBeeAddress64 addr = rx.getRemoteAddress64();
    if (addr == 0x0013A20040DADEE0 && type == 1 && b.len() == 8) {
        publish(F("Livingroom/Temperature"), b.remove<float>());
        publish(F("Livingroom/Humidity"), b.remove<float>());
        return;
    }
    if (addr == 0x0013A20040E2C832 && type == 1 && b.len() == 8) {
        publish(F("Study/Temperature"), b.remove<float>());
        publish(F("Study/Humidity"), b.remove<float>());
        return;
    }
    DebugSerial.println(F("Unknown or invalid packet"));
    printResponse(rx, DebugSerial);
}
```

Instead of simply printing the packet contents as before, it figures out who the sender of the packet is and which Beebotte resource corresponds to that, and calls the `publish()` function defined earlier.

As you can see, the Beebotte resources are identified using `Channel/Resource`, resulting in a unique identifier for each resource (it is later used in the MQTT message as the **topic identifier**). Also, note that the `F()` macro is used for the resource names to store them in program memory, as that is what `publish()` and the MQTT library expect.

If you run the resulting sketch, and everything connects correctly, the coordinator will forward any sensor values it receives to the Beebotte server. If you wait for (at most) five minutes to pass (or reset the sensor Arduino to have it send a reading right away) and then go to the appropriate channel in your Beebotte control panel, it should look something like this:

Configured resources		
Temperature *Ambient temperature*	21.70°	*3 minutes ago*
Humidity *Ambient Humidity*	62.40 %	*3 minutes ago*

Visualizing your data

To easily allow visualizing your data, Beebotte supports **dashboards**. A dashboard is essentially a Web page where you can add graphs, gauges, tables, buttons, and so on (collectively called **widgets**). These widgets can then display, or control, the data in one or more previously defined resources.

To create such a dashboard, head over to the **My Dashboards** section of your control panel, and click **Create Dashboard** to start building one.

Once you set a name for the dashboard, you can start adding widgets to it. To display the temperature and humidity for all the sensors you are using, you could use the **Multi-line Chart** widget. Since the temperature and humidity values will be fairly far apart, it makes sense to put them each in a separate chart. Adding the temperature chart could look like this:

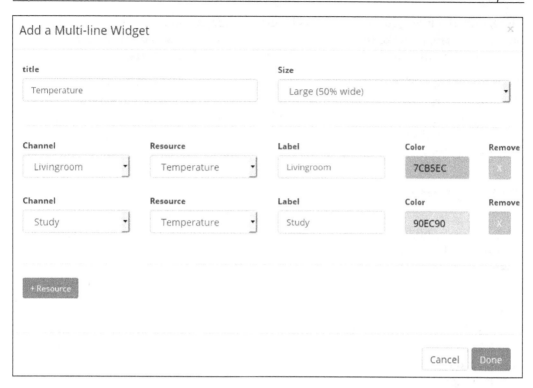

If you also add a chart for the humidity, it should look like this:

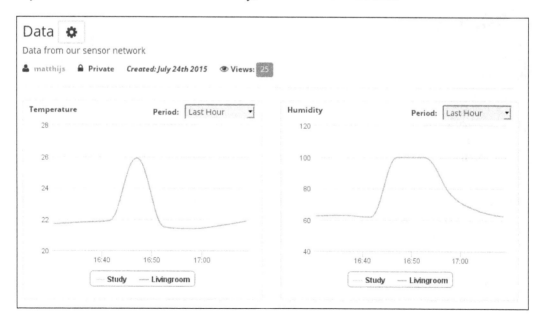

Here, only the living room sensor has been powered on, so no data is shown for the study yet. Also, to make the graphs a bit more interesting, some warm and humid air was breathed onto the sensor, causing the big spike in both charts.

There are plenty of other widget types that will prove useful to you. The Beebotte documentation provides info on the supported types, but you are encouraged to just play with the widgets a bit to see what they can add to your project. In the next chapter, you will see how to use control widgets, which allow sending events and data back to the coordinator to control it.

Accessing your data

You have been accessing your data through a Beebotte dashboard so far. However, when these dashboards are not powerful enough for your needs, or you want to access your data from within an existing application, you can also access recent and historical data through Beebotte's HTTP or WebSocket API's. This gives you full control over the processing and display of the data, without being limited in any way by what Beebotte offers in its dashboards.

Since creating a custom (Web) application is beyond the scope of this book, the HTTP and WebSocket API will not be discussed in detail. Instead, you should be able to find extensive documentation on this API on the Beebotte site at `https://beebotte.com/overview`.

Keeping your data locally

Instead of connecting your Arduino to the Internet and storing your data there, perhaps you want to maximize your control over your data and instead store it on your own computer.

There are plenty of options for doing so. Examples include: running a server in your local network that uses MQTT or some other protocol and having the coordinator connect to it, adding an SD card (Secure Digital) to the Arduino and logging data on there, or connecting the coordinator to a computer directly through USB and collecting the data on the computer, and so on.

Since creating the necessary scripts or setting up the right software is more work and more complicated than using a pre-existing online platform and there are so many options available, these will not be covered in detail in this book.

Instead, a single example will be given in which the coordinator will send its data through the USB serial connection to a computer, which will store the collected measurements in an SQLite database and show a simple graph with the most recent values.

Setting this up needs two parts: a sketch on the coordinator to send the data through USB, and a script running on the computer that receives and processes the data.

Both parts will be discussed in turn. For the computer side of things, the discussion will focus on how to get the provided example working, without discussing how it works in detail.

Sending data over the serial port

To get your measured data from the coordinator to a computer, the easiest approach is to send it through the virtual serial port created through the Arduino's USB connection. So far, you have been using this serial port to show debugging info on your screen, but this serial port can also be opened and processed by a program or script running on the computer.

The sketch to achieve this is fairly similar to the previous sketch, except that the publish() function needs to write to the serial port instead of sending things to the network. This is shown next:

```
void publish(const __FlashStringHelper *resource, float value) {
  DebugSerial.print(F("DATA:"));
  DebugSerial.print(resource);
  DebugSerial.print(F(":"));
  DebugSerial.println(value);
}
```

Whenever this function is called, it simply writes the resource name and its value to the serial port. The line it writes looks like this:

```
DATA:Livingroom/Temperature:21.60
```

The DATA: prefix allows the receiver to distinguish lines containing data from lines containing debug information. It is followed by the resource name and the value, separated by a colon.

There are certainly more robust and elegant ways of transmitting data, but this approach is simple and works well enough (provided that no colons are used in the resource names, and debugging output never starts with DATA:).

The complete sketch for this can be found as Coordinator_Serial.ino in the code bundle. If you run it, you can easily test it using the serial monitor. You should get lines similar to the one shown previously, intermixed with any debug output.

Receiving data over the serial port

On the computer on the other end of the USB connection, you will need to run a program to receive and process the data being sent by the coordinator. In this example, the Python programming language is used (see `https://www.python.org/`), since this is a powerful language, but also fairly easy to learn.

 Another tool often used to talk to an Arduino from a computer is Processing (`https://processing.org/`). Processing is especially suited for making interactive user interfaces and doing various data processing tasks, but less useful to run unattended in the background.

In the code bundle for this book, there is a Python script called `store_and_plot.py`, which can read the values from the serial port, store them in a database, and plot them in a graph on the screen.

To run this script, you will need:

- The Python programming environment, version 2.7 or higher (3.x is recommended)
- The `pyserial`, `matplotlib`, and `tzlocal` Python libraries

Getting these slightly depends on your platform:

- For Windows, Anaconda is a complete Python distribution that includes Python itself, the "pip" package installer, and a lot of libraries preinstalled (such as matplotlib), and can be found at `http://continuum.io/downloads`.

  After installing Anaconda, installing the other libraries can be done using:

  ```
  pip install pyserial tzlocal.
  ```

- On OSX, Python is provided by default but the Python package manager, `pip` is not. For this reason, it is recommended to install Python separately using `homebrew` (which can be found at `http://brew.sh/`). After installing homebrew, and Python using **homebrew** (see `http://docs.python-guide.org/en/latest/starting/install/osx/`), install the rest of the packages using:

  ```
  pip install matplotlib pyserial tzlocal
  ```

- On Linux, Python is typically provided by the distribution's package manager. Exact instructions vary by distribution but, on Debian or Debian-derived distributions (like Ubuntu), this would be a matter of installing the `python3`, `python3-pip`, `python3-matplotlib`, and `python3-serial` packages. The `tzlocal` library is not available through a Debian package and so must be installed through pip:

```
pip install tzlocal
```

After installing these dependencies, edit the `store_and_plot.py` script and modify the `SERIAL` variable with the name of the serial port offered by your Arduino (this is the same value you select under **Port** in the Arduino IDE when uploading to the Arduino). Make sure you have the serial console in the Arduino IDE closed, to prevent a conflict when accessing the serial port. Then run the script by opening up a console window, navigating to the directory that contains the script, and then running:

```
python3 store_and_plot.py
```

You might need to replace `python3` by `python` if you installed Python version 2.x.

Running the script should immediately open up a window showing a graph and waiting for any sensor readings to be received. All messages received through the serial port will also be printed in the console window. After some data has been received, a graph should be drawn:

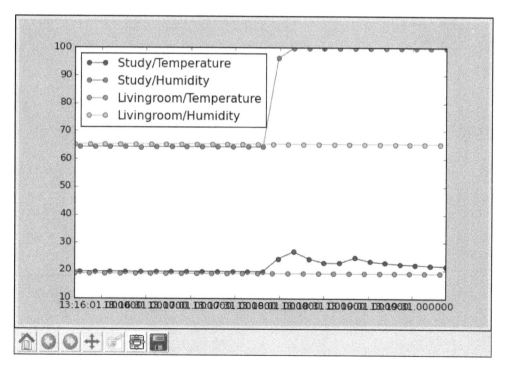

Discussing the workings of this Python script is beyond the scope of this book, but you are encouraged to further explore the Python programming language and use this script as a starting point if it fits your project. A good resource here would be *Python Programming for Arduino*, written by *Pratik Desai* (`https://www.packtpub.com/application-development/python-programming-arduino`).

In this example, we have talked about running the script on your computer, but you could also consider running a similar script on a dedicated mini-computer (such as the popular Raspberry Pi, which can run Linux). Without a screen attached, you will, of course, have to output the data in some other way (such as through a website, or by forwarding it to some other server).

Summary

In this chapter, you have seen how to get the sensor readings off your coordinator and into the cloud or onto your computer. You can visualize your data in convenient graphs, which gives you a great insight into your environment.

In the next chapter, you will continue building on top of this by letting your system not only monitor your surroundings, but also automatically control your heating or air conditioning system, based on the measurements made and the parameters you specify.

4

Controlling the World

By now, you have seen your Arduinos read their sensors, collect that data in the coordinator, and even store it in the cloud or on your computer. In this chapter, you will make your network take control of part of your surroundings.

The goal of this chapter is turn your network into a thermostat that can control the heating or cooling of (part of) your house. For this, there are two elements that need to be added:

- Some way to configure the intended temperature (setpoint)
- Some way to turn the heating and cooling on or off

The setpoint will be controlled through the Beebotte dashboard, allowing you to control your house's temperature from any browser (including one on a smartphone). If you prefer to build an actual thermostat device using a display and a knob, that is also possible (you can even combine both), but this is beyond the scope of this book.

Controlling the heating or cooling is a bit more tricky and depends on what kind of system you have. Some options will be suggested using a remotely controllable power outlet or a relay module (Arduino-controlled switch).

For controlling the wireless power outlet, you will dive into the ZigBee and ZigBee Home Automation specifications a bit to figure out how to talk to an off-the-shelf wireless power plug. This should give you a head start for working with other ZigBee devices as well.

This chapter has no list of recommended hardware, since the hardware to use greatly depends on what you want to control exactly, and what devices are available to you. Instead, some recommendations are made in the chapter (in terms of specific hardware items or generic kinds of hardware).

Controlling your heating and/or cooling system

To make a working thermostat, you will need some way to let it control the temperature in your house. One way to approach this is to let an Arduino replace your existing thermostat and have it control the HVAC (heating, ventilation and/ or air conditioning) system in your house in the same way the thermostat did. Alternatively, you could have your network control mains power to a device, such as an electrical heater or standalone air conditioning unit, and control the room temperature through that.

Ideally, you could toggle heating or cooling for each room in your house separately. Combined with a temperature sensor in each room, you would have detailed control over the temperature in every room of your house.

For simplicity, these examples assume that you will control the heating using a single on/off switch for the entire house. If you (instead or additionally) want to control cooling, the examples will be easy to modify. If you have a way to control rooms separately, modifying the coordinator sketch to support this involves a bit more work, but it should certainly be possible as well.

These examples are aimed at controlling a heating system, but it is just a matter of changing some of the naming and reversing the decision process to make it control cooling instead.

Replacing the thermostat

The simplest heating systems have two wires running from the boiler to the thermostat; the thermostat then connects them when there is a heat demand and disconnects again when there is not. Thermostats connected like this are typically referred to as **on/off thermostats**. Some air conditioning and ventilation systems work in a similar way; full HVAC units often provide multiple pairs of wires.

To control such a system using an Arduino, you need some way to connect two wires together. The easiest way to do this is to use a **relay**. This is a device containing a switch and an electromagnet. When power is applied to the magnet, it pulls the switch closed. Since there is no electrical connection between the magnet and the switch, the controlling system (the Arduino) and the switched system (the HVAC system) are isolated.

Some, especially older HVAC systems apply the mains voltage on their on/off switching pins, essentially running the full electrical load through the thermostat. See the following remarks about controlling mains power, which apply here as well.

The easiest way to get a relay attached to your Arduino is to use a relay shield, such as the one sold by SparkFun (`https://www.sparkfun.com/products/12093`). Using a shield, all the necessary connections have been made and you can switch the relay on by writing `HIGH` to the relay control pin on the Arduino. Relay modules (such as `https://www.sparkfuns.com/products/11042`) are an alternative; they have the same components to directly control the relay, but do not use the Arduino shield form factor.

Modern HVAC systems often additionally offer **modulating** control for more advanced thermostats. Instead of just turning the heating on or off, they can control the boiler power and, for example, set it to run at 30 percent power continuously. This kind of control is beyond the scope of this book, but often these systems also support on/off control and can still be used with these examples. Check the documentation for your HVAC system to be sure.

Controlling mains power

Controlling a smaller heating or cooling device (such as a standalone electrical heater, fan, or air conditioning unit), can be easiest by controlling the mains power to it. This works well for simple devices: apply the power and they turn on, cut power and they turn off. This might not be appropriate for all devices and will likely cause problems with a house-wide HVAC system, so use common sense here.

To control mains power, you would typically also use a relay. However, because working with mains power poses significant a health and fire hazard, it is not recommended to wire up a relay to mains power yourself unless you already know all the constraints and caveats involved.

Better would be to use an off-the-shelf device containing a relay, such as the **PowerSwitch Tail** device (see `http://www.powerswitchtail.com/`). This is similar to a relay module, but it has the relay safely isolated in an enclosure, offering an external connection for the Arduino directly (only needing GND and an I/O pin on the Arduino to connect to the **-in** and **+in** on the PowerSwitch tail side). From the Arduino, it looks just like a relay and can be controlled in the same way. Make sure you get the PowerSwitch Tail and not the PowerSSR tail if you need to switch big loads, since the latter has modest current capability.

An attractive alternative can be a power socket that supports the ZigBee wireless protocol. These do not require an additional Arduino and XBee module to control them, but can be controlled through your ZigBee network directly. The availability of these is limited, but the Meazon Izy plug can be recommended (as well as the Meazon Bizy, which also offers power usage measurements). See `https://izy.meazon.com/` for more information. If you are looking for an alternative, go ahead and read this chapter first to get a better idea of the possibilities and constraints with regard to the ZigBee protocol support.

Hairdryer – an alternative

If none of these options allow you to control your heating and/or cooling system, but you still want to work through this chapter, consider using a very simple heating device: a hairdryer. If you get a Meazon Izy or similar plug, connect the hairdryer to it, and put one of the temperature sensors in the range of the hairdryer; you can see how your network will use the hairdryer to stay warm (this does need the temperature sensor to send updates a lot faster, since the hairdryer might melt the sensor if it stays on for 5 minutes).

 Be careful when you run your system unattended, especially with devices such as electrical heaters or hairdryers, which can cause a fire hazard if flammable materials are present.

Control systems

A thermostat, as well as the system you are about to implement is an example of a **control system**. Simply put, a control system is a system that, based on a number of inputs, continuously decides on a value for one or more outputs. Most control systems are **closed loop control systems** (also known as **feedback control systems**), meaning that their outputs also influence the inputs, providing the system with feedback about the decisions it makes.

In this case, the inputs are the current temperatures and the setpoint temperature, and the single output is whether the heating system should be on or off. The feedback happens when the temperature changes due to the heating being on or off.

A lot of different sorts of control systems have been designed and documented but, like most on/off thermostats, this example will use a simple **hysteresis controller**. Consider the following approach: If any rooms have a temperature below the setpoint, turn the heating on; if all rooms are above the setpoint, turn the heating off.

This is simple enough and will work, but has one important problem: when the coldest room is at the setpoint, the heating will be rapidly toggling between on and off. Whenever the temperature drops by 0.1°, the heating is turned on and, as soon as the room starts heating up slightly, the heating is turned off again. While this will result in a temperature very close to the setpoint, it will typically be very inefficient and potentially cause unneeded wear in the HVAC system and relay.

The typical approach to solving this is to apply **hysteresis**: slightly modify the setpoint depending on whether the heating is currently on or off. For example, turn the heating on when any temperature is below the setpoint, but do not turn it off again until all temperatures are 0.5° above the setpoint. This ensures that when the heating is turned on, it is always left on long enough to heat up the room by half a degree and it is left off long enough for that half degree to cool again. This results in fewer changes in the heating state, but also a less precise system: the hysteresis creates a range of temperatures near the setpoint that are now acceptable, instead of just the setpoint value itself.

More advanced approaches exist, typically taking into account the history of inputs and outputs, or making a prediction of future inputs. For example, thermostats typically turn off the heating a bit before the setpoint is reached, anticipating that the temperature will rise a bit more because the radiators are still warm (this effect is called **overshoot**).

When the outputs become more complex than a simple on/off control, different kinds of controller are used. A very common controller is the PID-controller (proportional-integral-derivative), which uses a mathematical approach to decide the output based on the historical, current, and future values of the inputs.

With that bit of theory out of the way, it is time to get started. First, you will add setpoint control to your Beebotte dashboard, add the thermostat logic to the coordinator, and have it print its intended heating state to serial. Then you will explore a few ways to actually control the heating: using a relay attached to the coordinator directly, using a relay attached to another Arduino, and using a ZigBee Home Automation on/off power outlet.

Adding setpoint control

As a first step to turn your coordinator into a thermostat, you will need some way to control the setpoints. In this example, a single setpoint for the entire house is used, but it should be easy to add one per room and adapt the code to check each room temperature against the appropriate setpoint.

To store the setpoint, create a new channel in Beebotte called House. Inside, add a Setpoint resource, using the temperature type from the dropdown.

For this resource, you should enable the **Send on Subscribe (SoS)** option. Normally, when you subscribe to a resource, you will only receive new values published to it. For the setpoint, this will mean that the coordinator will get notified whenever the setpoint is changed. However, when the coordinator Arduino is first turned on or resets, it will not know about the current setpoint until it changes. The SoS option solves this by automatically sending the current value of the setpoint whenever the coordinator subscribes to it, so it will always have access to the most recent value.

To allow changing the resource, edit your dashboard and add an **Input** widget:

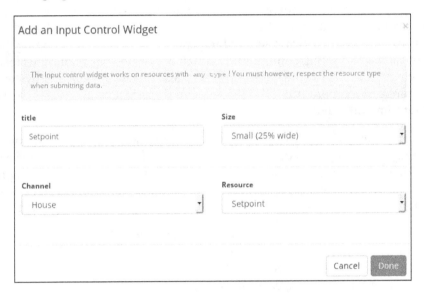

This should give you an input box on your dashboard, where you can input a setpoint temperature and click **Send**. This should look something like this:

 When this book was written, control widgets had only very recently been added to Beebotte and were still considered experimental. When you read this, it is likely that some details about them will have changed and more advanced control widgets will have been added, so be sure to check out the Beebotte documentation at `https://beebotte.com/overview`.

Note that the value you enter will be stored inside Beebotte, but it is not actually sent to your coordinator yet, since it did not **subscribe** to the House/Setpoint resource. The Coordinator.ino sketch in the code bundle for this chapter shows how to do this. It is based on the Coordinator.ino sketch from the previous chapter, but with significant changes and additions.

First off, you will need some place to let the coordinator store the current setpoint. This is a simple global variable called setpoint:

```
float setpoint = NAN;
```

Floating point variables can have a special value called **not a number** (NaN, or NAN in your code). This value is commonly used to indicate errors or other exceptional circumstances; here it is used to indicate "no setpoint received yet". NaN is interesting, because all calculations involving NaN will also result in NaN and comparisons involving NaN will always be false. These properties will come in handy when implementing the actual controller later.

There is another global variable you will need: inputs_changed:

```
bool inputs_changed = false;
```

This variable will be set by processRxPacket() when any of the room temperatures change, so the loop() function knows it should re-evaluate the current heating state.

Instead of setting a variable like this, you could also just update the output immediately when any temperature is changed. However, when updating the output involves sending a message through the XBee module and you want to wait for the transmit status value, this cannot happen inside an XBee callback. Since the reception of new temperature values from a sensor node happens inside the processRxPacket() callback, updating the output directly will cause problems. By setting a variable inside the callback and, later, outside the callback, checking that variable, and updating the outputs, this problem is avoided.

Subscribing to events

To make your Arduino get notified whenever you change the setpoint, it needs to subscribe to the resource through the MQTT protocol. After subscribing, the MQTT server will send any new values for the subscribed resource to the Arduino, which can then process them appropriately.

To subscribe using the Adafruit MQTT library, you will need a global variable of the Adafruit_MQTT_Subscribe type. This variable is a composite object that will track the topic subscribed to, as well as the last value received. In your code, this will look like this:

```
const char HOUSE_SETPOINT[] PROGMEM = "House/Setpoint";
Adafruit_MQTT_Subscribe setpoint_subscription(&mqtt, HOUSE_SETPOINT);
```

Here, &mqtt refers to the global mqtt object created before. The MQTT topic to subscribe to (which refers to the Setpoint resource in the House channel in Beebotte) must be stored in PROGMEM, just like the MQTT server connection details you saw before.

Now, this only prepares the subscription. To actually subscribe to the topic, you will need to pass this object to the mqtt.subscribe() method:

```
mqtt.subscribe(&setpoint_subscription);
connect();
```

Subscribing should happen in your setup() function, but before the call to the connect() function. At the time of writing, the Adafruit MQTT library does not support adding subscriptions after a connection is made, even though the MQTT protocol supports this just fine.

Reading events

Now that you have subscribed to the setpoint resource, the MQTT server will send the values to your Arduino. To actually read and process them, you will need to add a bit of code to your loop() function:

```
Adafruit_MQTT_Subscribe *subscription = mqtt.readSubscription(0);
if (subscription)
  handleSubscription(subscription);
```

This code uses the mqtt.readSubscription() method, which checks if there is a new message from the MQTT server waiting and, if so, processes it and returns a pointer to the subscription that received a new value (returning NULL if no message was received). The single argument to mqtt.readSubscription() is how many milliseconds it should wait for new data to arrive before returning NULL. In this case, 0 is passed, so it just checks if anything is pending and, if not, returns immediately again (no need to wait, since the method will be quickly called again anyway).

If a value was indeed received, mqtt.readSubscription() returns a pointer to the relevant Adafruit_MQTT_Subscribe object. In this case, you will know it is the setpoint value, but if there are multiple subscriptions, this pointer lets you see what subscription produced a new value.

The actual handling of the value is forwarded to the `handleSubscription()` function, which looks like this:

```
void handleSubscription(Adafruit_MQTT_Subscribe *subscription) {
  float value = getFloatValue(subscription);
  DebugSerial.print((const __FlashStringHelper*)subscription->topic);
  DebugSerial.print(F(" changed to "));
  DebugSerial.println(value);

  if (subscription == &setpoint_subscription) {
    setpoint = value;
    inputs_changed = true;
  }
}
```

The value received will be JSON-encoded by Beebotte, just like the values you publish to them, so it is first converted to a floating point value and printed. Then the subscription is checked (just in case, and to provide an example if you want to add more subscriptions) and the global `setpoint` variable is updated. Finally, `inputs_changed` is updated, so the output can be updated if needed (as shown later).

The full JSON value received through MQTT looks like this:

```
{"data":25.2,"ts":1438674255995,"ispublic":false}
```

The function to extract the `25.2` part from this and convert it to a floating point value looks like this:

```
float getFloatValue(Adafruit_MQTT_Subscribe *subscription) {
  String str((const char*)subscription->lastread);
  const char *PREFIX = "{\"data\":";
  // Check for the prefix
  if (!str.startsWith(PREFIX)) {
    DebugSerial.print(F("Unsupported value received: "));
    DebugSerial.println(str);
    return NAN;
  }
  // Remove the prefix
  str.remove(0, strlen(PREFIX));
  // Convert the rest into float (as much as possible)
  return str.toFloat();
}
```

This is not the most elegant way to parse a bit of JSON (if the ordering of fields changes, the code needs to be changed as well, for example), but it works well enough for this situation. This code converts the `subscription->lastread` buffer, which contains the received data, into a `String` object for easier processing (a cast to `const char*` is needed because `lastread` is an array of `uint8_t` and `String` only accepts `char`). Then it confirms that the received value starts with the expected prefix and strips the prefix, making the number appear at the start of the string. Finally, `toFloat()` takes care of converting the number to a proper floating point value, ignoring any extra characters after the number.

> The Adafruit MQTT library truncates the value during reception (to 19 bytes, by default). Since this only cuts off unused data here, this is no problem, but keep this in mind if you need to receive longer values. This cutoff length can be changed through `SUBSCRIPTIONDATALEN` in `Adafruit_MQTT.h`.

If you upload the sketch as it is now, you should be able to change the setpoint in Beebotte and have the change show up on your serial monitor. This could look something like this:

```
Starting...
Ethernet initialized
MQTT connected
House/Setpoint changed to 10.00
House/Setpoint changed to 21.00
```

Remembering sensor readings

With the setpoint input to your thermostat covered, it is time to look at the rest of the inputs: the measured temperatures. Right now, any sensor readings received are forwarded to Beebotte, but not remembered by the coordinator. If these values are to be used to control the heating, they must be stored. For this, we need an array of temperature values, one for each room. To help identify the rooms, first add a global enumeration:

```
enum {
    LIVINGROOM,
    STUDY,
    NUM_ROOMS,
};
```

This `enum` essentially defines three constants, with consecutive values (0, 1, and 2). If you make sure that `NUM_ROOMS` is always the last element in the list, this easily lets you add new rooms, giving them a unique index automatically, while at the same time keeping `NUM_ROOMS` accurate.

With these defined, you can add the `temperatures` array:

```
float temperatures[NUM_ROOMS] = {NAN, NAN};
```

Values are initialized to NaN, just like the `setpoint` variable, to indicate that no valid value is known yet.

Finally, values must be assigned to this array whenever they are received. Where `processRxPacket()` previously only called `publish()` for any received values, it should now also update the `temperatures` array. Here is the relevant piece of `processRxPacket()` with that change applied:

```
if (addr == 0x0013A20040DADEE0 && type == 1 && b.len() == 8) {
    temperatures[LIVINGROOM] = b.remove<float>();
    publish(F("Livingroom/Temperature"), temperatures[LIVINGROOM]);
    publish(F("Livingroom/Humidity"), b.remove<float>());
    inputs_changed = true;
    return;
}
```

Note that this first stores the temperature value to the `temperatures` array and then passes that value to `publish()`. Using a call to `b.remove<float>()` in both these places would not work here; the first call will retrieve the temperature and remove it from the buffer, the second call will return the humidity instead.

This also sets the `inputs_changed` global variable to make sure the output is updated if needed. If you have multiple sensor nodes and a block of code for each, all of them should be updated, of course.

Thermostat controller

Now all the inputs are covered, it is time to implement the actual control system that looks at the inputs and decides on the output. Since this decision is influenced by the current heating state (this is the hysteresis mechanism), you will need to keep track of the current state, too, necessitating another global variable:

```
bool heating_state = false;
```

The decideHeatingState() function will be the core of the control system and is called whenever an input is changed. It should return whether the heating should be on or off:

```
bool decideHeatingState() {
  // Apply hysteresis
  float corrected = heating_state ? (setpoint + 0.5) : setpoint;
  for (uint8_t i = 0; i < NUM_ROOMS; i++) {
    // At least one room is cold, heating should be on
    if (temperatures[i] < corrected)
      return true;
  }
  // All rooms are warm enough, heating should be off
  return false;
}
```

This implements exactly the control strategy outlined earlier: when any room's temperature is below the setpoint, turn the heating on, and when all rooms have their temperature 0.5° above the setpoint, turn it off again.

The hysteresis is implemented by artificially moving the setpoint up by half a degree when the heating is currently on. You might need to tweak this hysteresis offset a bit, depending on the results.

Remember that a comparison involving NaN will always return false. Since the setpoint and temperature values are initialized to NaN, the heating will not turn on unless a setpoint and at least one room temperature have actually been received (so this property of NaN values removes the need of an explicit check to this effect).

Now that you have the decision-making-function, you must actually call the function whenever an input has changed and make sure that its decision is executed. The following addition to the loop() function will take care of that:

```
if (inputs_changed) {
  if (decideHeatingState() != heating_state)
    switchHeating(!heating_state);
  inputs_changed = false;
}
```

This calls the switchHeating() function to actually change to the new heating state. The single parameter passed is the new state. Note that the preceding code does not actually change the heating_state variable yet. Instead, it expects the switchHeating() function to update heating_state, but only if the switch actually succeeds. If a failure occurs (such as a transmission failure), heating_state will keep its old value (correctly reflecting reality) and another attempt at switching the heating will be made when the inputs change next.

For the actual `switchHeating()` function, try this dummy version first:

```
void switchHeating(bool state) {
  Serial.print(F("Heating should be "));
  Serial.println(state ? F("on") : F("off"));
  heating_state = state;
  publish(F("House/Heating"), state);
}
```

This will not actually switch the heating, but will just print a debug message, update the `heating_state` variable, and publish a message to Beebotte. The latter requires adding a new `Heating` resource in the `House` channel, which should have a `boolean` type (since it can only have two values: true/on or false/off).

With one last addition, this version of the sketch is complete. Add this line at the end of the `setup()` function:

```
switchHeating(false);
```

This ensures that the heating is switched off at startup. This is not strictly necessary, but can be helpful as a safety measure. If the Arduino is ever unintentionally reset when the heating is on, this will make sure that the `heating_state` variable still reflects reality, preventing the heating from staying enabled when it is not needed.

The sketch described up to now is provided in the code bundle as `Coordinator. ino`. Go ahead and upload it to see if your control system behaves as expected. Since there is no feedback yet, you will need to play with the setpoint (or temperatures — manually turning a hair dryer on and off could help) to see the decision actually change.

In the next sections, a couple of different versions of the `switchHeating()` function will be shown: using a relay attached to the coordinator directly, using a relay attached to a remote Arduino, or using an off-the-shelf ZigBee power plug. In *Chapter 5, Standalone XBee Operation*, one more version will be shown, to control a relay directly attached to an XBee module.

Controlling a relay

Controlling a relay module or shield is fairly simple: Connect it to an Arduino using 5V, GND and a single I/O pin. Then write HIGH to that I/O pin to turn the relay on, write LOW to switch it off again. Using the PowerSwitch Tail is identical, except you only connect GND and an I/O pin; the Tail has its own power supply built in.

In these examples, we will assume the relay is controlled through pin 4 (as used by the SparkFun shield) and use a constant to store the pin number:

```
const uint8_t RELAY_PIN = 4;
```

Of course, the pin must be configured as an output in the `setup()` function:

```
pinMode(RELAY_PIN, OUTPUT);
```

Now, if you can attach the relay directly to the coordinator, the `switchHeating()` function will be rather trivial:

```
void switchHeating(bool state) {
  digitalWrite(RELAY_PIN, state);
  heating_state = state;
  publish(F("House/Heating"), state);
}
```

This version of the sketch is available in the code bundle as `Coordinator_Relay.ino`.

Things become a little more complicated if the relay is not attached to the coordinator directly, but to another Arduino. This can be an existing Arduino (with a temperature sensor, for example), but this examples assumes that there is a dedicated Arduino (with XBee module) just for controlling the relay.

 It is also possible to let an XBee module control a relay directly, without having an Arduino connected. *Chapter 5, Standalone XBee Operation*, shows how to implement this.

Since the coordinator is the one that makes the decisions, there must be a way for the coordinator to instruct the relay Arduino to toggle the relay.

For this, it will send a message through the XBee module. In *Chapter 2, Collecting Sensor Data*, you defined a *Temperature and humidity data* message format, identified by an initial 0x01 byte. Now you will add a new *Switch relay* message format identified by an initial 0x02 byte. To know what the state of the relay should be, the message needs a second byte, which will be 0x00 when the relay is off, and 0x01 when the relay is on. This packet format can be summarized like this:

0	1
type	on/off
(2)	(0/1)

To send these messages, the `switchHeating()` function in your coordinator can be defined like this:

```
void switchHeating(bool state) {
  // Build the packet
  AllocBuffer<2> packet;
  packet.append<uint8_t>(2); // Packet type: Switch relay
  packet.append<uint8_t>(state ? 1 : 0); // Value
  // Send the packet to the node with the relay
  XBeeAddress64 addr(0x0013A20040E2C832);
  ZBTxRequest txRequest(addr, packet.head, packet.len());
  if (xbee.sendAndWait(txRequest, 5000) == 0) {
    heating_state = state;
    publish(F("House/Heating"), state);
  } else {
    DebugSerial.println(F("Failed to send packet to relay"));
  }
}
```

This is very similar to the `sendPacket()` function in your `DhtSend.ino` sketch from *Chapter 2, Collecting Sensor Data*, except that it sends different data in the message. As noted before, `heating_state` is only updated when the heating was successfully switched by checking the transmit status returned by the XBee module. A zero status means the packet was acknowledged by the remote XBee module, even though there is no guarantee that the remote Arduino actually processed it correctly.

Do not forget to use the address of your relay-controlling Arduino here. This version of the sketch is included in the code bundle as `Coordinator_Remote_Relay.ino`.

At the relay Arduino, these packets must be received and acted upon. The sketch for this, called `Relay.ino` in the code bundle, is very similar to the `Coordinator.ino` you built in Chapter 2 (before adding any of the Internet connectivity), but with a different `processRxPacket()` function:

```
void processRxPacket(ZBRxResponse& rx, uintptr_t) {
  Buffer b(rx.getData(), rx.getDataLength());

  uint8_t type = b.remove<uint8_t>();
  // Packet type 2 is "Switch relay" command
  if (type == 2 && b.len() == 1) {
    uint8_t state = b.remove<uint8_t>();
    digitalWrite(RELAY_PIN, state);
    DebugSerial.print(F("New relay state: "));
    DebugSerial.println(state);
  } else {
    DebugSerial.println(F("Unknown packet type, or invalid length"));
  }
}
```

Do not forget to also add the `RELAY_PIN` constant and `pinMode()` call shown earlier as well. With this sketch, you should be able to control the relay (or PowerSwitch Tail) from your coordinator and remotely switch on your heating when needed.

This setup using a relay can be useful to control all kinds of other devices as well (but remember not to mess around with mains power!). Additionally, this example can be easily adapted to control more complex devices, such as a (servo) motor, display, infrared transmitter, and so on. The basic structure will be the same, but the data transmitted will be different.

Controlling off-the-shelf ZigBee devices

Instead of attaching things to an Arduino to expand the input and output capabilities of your network, you could also consider off-the-shelf ZigBee devices. In particular, devices that implement the ZigBee Home Automation specification (covered in more detail next) can typically add useful things, such as on/off control of power outlets, (wall) switches or dials, displays, various sensors, and so on.

In this section, you will see how to remotely control a power outlet that supports the ZigBee Home Automation standard, making it turn on and off. Since the protocols and specifications needed for this are complex and big, this will only cover the most basic use case. Implementing the complete ZigBee Home Automation specification using an Arduino and XBee module could probably fill a book on its own, but this section should give you enough info to allow basic control of an on/off device and give you a starting point to implement other things, as well.

To be able to talk to these other ZigBee devices, you will have to know what message to send to them so that they will respond appropriately. So far, you have been sending data from one XBee module to another. A message consisted of a single sequence of bytes and it was up to you how to interpret them. These messages used a type of ZigBee packet that is specific to the XBee modules, but ZigBee defines a lot of other packet types too.

To understand the kinds of packet that are available within the ZigBee protocol, you will first need to learn about **profiles**, **endpoints**, and **clusters**.

 This section applies exclusively to XBee ZB modules, using the ZigBee protocol. No equivalent standards exist for the other XBee families.

ZigBee profiles, endpoints, and clusters

The ZigBee protocol itself defines the general networking infrastructure to allow joining a network, discovering devices and services on those devices, and of course, sending and receiving data. However, the ZigBee protocol does not actually define what kind of messages devices should actually exchange to get something done. The latter is done by various ZigBee **application profiles**. There are a number of public ZigBee profiles, each of which defines a number of device types (such as light bulbs, smart meters, wall switches, heaters, and so on). For each device type, a profile specifies what messages such a device should be able to send and receive and how it should behave.

To allow a device to support multiple examples of those device types, a ZigBee has one or more **endpoints**. Every endpoint gets exactly one device type assigned to it, so a device will have one active endpoint for every device type it supports. Every device can support up to 240 endpoints, numbered from 1 to 240 (identifiers are arbitrarily selected by the device manufacturer). These can be device types from different profiles, multiple device types from the same profile, or even the same device type multiple times.

Endpoint 0 is reserved for the **ZigBee Device Profile** (**ZDP**). This is a mandatory profile (all ZigBee devices must support it) that is used for discovering devices and their capabilities and for some configuration and network management tasks. Later, you will use these ZDP commands to detect the available endpoints on a ZigBee device.

Each profile specifies the functionality for each of its device types, but not directly as a big chunk of requirements. Instead, the functionality is divided into different **clusters**. Each cluster is essentially a bit of related functionality, and typically defines some commands that can be sent or received, and/or attributes that can be read or written.

Multiple device types can use the same clusters, preventing duplication in the specifications, and a codebase supporting multiple device types. For example, the ZigBee Home Automation profile defines the mains power outlet and heating/cooling unit device types. These are clearly different devices with different requirements, but both of them support being turned on and off. To support this, both implement the *on/off* cluster, which specifies commands to turn a device on or off, and an attribute to read the current state.

To further encourage sharing of clusters, even between different profiles, the ZigBee alliance published another specification called the **ZigBee Cluster Library** (**ZCL**). This specification defines a large number clusters, for various purposes, that are used by all the profiles from the ZigBee Alliance.

Every cluster is further subdivided into a **server** side (also called **input cluster**) and **client** side (also called **output cluster**). Typically, communication is initiated by the client side sending a command to the server side, which then responds to the client side. Any endpoint can implement one or both sides of a cluster.

Every ZigBee radio message exchanged contains a source and destination endpoint ID, a profile ID, and a cluster ID that tell the receiver what the message is intended for and how to interpret the message content. Up until now you have been sending messages between your XBee modules. If you could sniff the actual radio packets out of the air, you would see that these too contain (XBee-specific) endpoint, profile and cluster identifiers automatically added by the XBee module.

Together, all this results in a system that can be used to specify functionality for a large number of different device types, with relatively little duplication in the specification. The flip side of this is that looking up what bytes you need to send to a device to make it turn on can be a challenging journey through two or three specification documents, each of them hundreds of pages long, which can be intimidating. As a guide to one such journey, this section will walk you through the process for getting your Arduino to connect to a ZigBee-enabled wireless power outlet and remotely turn it on and off.

ZigBee public profiles

The ZigBee Alliance has published a number of standardized profiles, referred to as **public profiles**. Their full specifications can be downloaded from http://www. zigbee.org free of charge and they can be used to talk to any device implementing any of these public profiles. At the time of writing, the public profiles were:

- **ZigBee Home Automation (ZHA)** and **ZigBee Building Automation (ZBA)**: These profiles are aimed at typical home automation devices such as lights, switches, power outlets, thermostats, and so on. They mostly specify the same device types, but ZHA is aimed at homes while ZBA is aimed at commercial buildings (and has a slightly more complex network-forming procedure).

- **ZigBee Health Care (ZHC)**: This profile allows communication between health care-related devices.

- **ZigBee Retail Services (ZRS)** and **ZigBee Telecom Services (ZTS)**: These two profiles are aimed at information exchange and payment services for retail stores, using special-purpose hardware (ZRS), or mobile devices such as smartphones (ZTS).

- **ZigBee Smart Energy (ZSE)** and **ZigBee Light Link (ZLL)**: These profiles are aimed at smart metering networks that are controlled by utility companies and remotely controlling light bulbs or other lighting-related devices. It would be great if you could interface with these kinds of devices, but unfortunately there are security measures in place that prevent this. Even though the protocol specification is open, you will need a secret key (ZLL) or certificate (ZSE) to communicate with these devices, which you can only obtain if you build a device and get it certified by the ZigBee alliance. Additionally, ZLL and ZSE sometimes use special **inter-pan** messages to talk to a device without joining their network first, which XBee modules do not currently support.

Some ZLL devices also implement ZHA, so you can still interface with them by letting them join your ZHA network.

There are some other standards such as ZigBee Input Device, ZigBee Remote Control, and ZigBee 2030.5 (also known as ZigBee Smart Energy 2.0). To add to the complexity and confusion, these standards do not build on top of the ZigBee PRO protocol (as used by the XBee ZB modules), but instead build upon the ZigBee RF4CE or ZigBee IP protocols, which are alternatives to ZigBee PRO.

Finally, there is ZigBee 3.0 (scheduled to be published by the end of 2015), which is intended to replace all of the preceding profiles in a single unified profile, in a backward-compatible way (though currently available documentation does not indicate how this compatibility will be achieved, and whether this could somehow lift the secret key limitations of ZLL and ZSE).

Selecting a ZigBee device

Using the applicable specifications listed earlier, it should be possible to take any off-the-shelf piece of electronics that supports one of these ZigBee profiles (except ZLL and ZSE as noted) and talk to it from your Arduino through an XBee module.

Even though ZigBee is a popular wireless technology, finding an example device to use in this book turned out to be harder than expected. A few of the problems encountered were:

- Product information often does not prominently indicate that ZigBee is used, or what ZigBee profile is used. The ZigBee Products page on http://www.zigbee.org can help here, though.

- Some devices advertising ZigBee actually use a (undocumented) private profile on top of ZigBee PRO instead of a public profile.

- Manufacturers often have interesting products, but no store can be found that actually sells them (perhaps because the products are new, or they only sell through installers).

- Products are typically localized (and available exclusively in the US, or in Europe, and so on), especially when they involve a power socket.

- Product names and codes are not always consistent, making it sometimes hard to know that a product in a store is really the same product you saw on `http://www.zigbee.org` or on the manufacturer's site.

After a bit of digging, the Meazon Izy on/off plug was found (see `https://izy.meazon.com/`). It is readily available from several global online stores and uses the ZigBee Home Automation profile. Unfortunately, it is only available with the "Schuko" type of socket (which is used in most of Europe). Trying to find an alternative for the US or UK markets was unsuccessful at the time of writing (due to the aforementioned problems).

In the following examples, this Meazon Izy plug is used:

It is a fairly simple device: you plug it into a wall socket and you plug another device into the Izy, and then you can remotely turn the power on and off through ZigBee commands. The Bizy plug by the same manufacturer can additionally do power usage measurements as well.

Often, ZigBee devices are sold in some kind of starter kit, containing a single gateway or coordinator device and one or more actually useful devices. When you want to integrate these into your own network, you typically do not need any gateway device, so make sure to look for **add-on packs** that come without such a gateway.

Talking to a ZigBee on/off device

Now that you have learned about some ZigBee terminology and available devices, it is time to start talking to one. Next, all the steps required to figure out how to send the on/off commands are laid out, citing the relevant specifications where appropriate. Hopefully, this will not only allow you to control a Meazon Izy plug, if you have one, but will also help you figure out any other ZigBee (Home Automation or other) devices that you want to use in your network.

The following specifications might prove useful during this process:

- The ZigBee specification, revision 20 (Document 053474r20, `http://www.zigbee.org/download/standards-zigbee-specification/`)

- The ZigBee Home Automation Public Application Profile, revision 29 (Document 05-3520-29, `http://www.zigbee.org/non-menu-pages/zigbee-home-automation-download/`). Next, this document is referenced using *ZHA* section *x.y*.

- The ZigBee Cluster Library specification, revision 20 (Document 075123r04ZB, `http://www.zigbee.org/download/standards-zigbee-cluster-library/`). Next, this document is referenced using *ZCL* section *x.y*.

Overall, these are the steps you will need to take:

1. Get the new device to join your network.

2. Send it ZDP service discovery commands to figure out what kind of endpoints, profiles, and clusters it supports.

3. Figure out what commands to send, and send them.

Joining the network

Typically, when a ZigBee device powers up, it scans for any available networks and tries to join one that it likes. Sometimes the device keeps doing this until it has joined and sometimes it requires a press on a button to even start in the first place.

 It is recommended to keep your ZHA device turned off until you have prepared your XBee network, to make sure it can join the network correctly on the first attempt.

Getting a ZHA device to join your network turns out to be fairly easy. There are two changes needed to the XBee configuration:

- Set ZS=2: ZHA uses the **ZigBee PRO** stack profile (see ZHA section 5.2). By default, XBee modules advertise a **network specific** stack profile, which needs to be changed to ZigBee PRO.

- Set KY=5A6967426565416C6C69616E63653039: When joining a network, ZHA devices use a default trust center link key to obtain the network key (see ZHA section 5.3.3). The key to use is listed in the ZHA specification and is shown here.

> This changes the trust center link key to a publicly known key, which as good as negates the security advantage of using this key in the first place. Be aware that this makes your network less secure while joining is enabled, so make sure you only enable network joining when strictly needed (and keep NJ=0 at other times). Refer back to the *Disabling network joining* section in *Chapter 1, A World without Wires*, for more details.

Changing ZS or KY will cause an XBee module to leave the current network and start (coordinator) or join (router/end device) a new network. Also, since a node cannot join a network if either the stack profile or link key is different from the coordinator, you must change both values on all XBee modules in your network.

Some additional information on how a ZHA device joins a network can be found in *ZHA* section, under *Network steering*. That same section talks about *EZ-Mode Finding & Binding* and *Centralized Commissioning* for connecting devices together (for example to tell an on/off switch what outlet it should switch), but you do not need that here; an outlet can be switched by anyone that sends it the right command, even it was not **bound** beforehand.

After you make the preceding configuration changes, temporarily enable network joining (using NJ or a commissioning button). Then, power up your ZHA device; it should find your network and join it right away. To confirm, you can use the network scan feature in XCTU, or the discovery sketch described next.

Factory reset in case of problems

When joining a device to your own network does not work, it might help to do a factory reset or push some button, forcing the device to re-attempt a join. A factory reset can also be needed to get a device to join another network (for example, after you changed some parameters that caused a new network to be created).

How this works exactly is different per device, but this typically means pressing or holding a button, holding a magnet near the device, repeatedly disconnecting and reconnecting power to the device, or a combination of these.

Not all manufacturers provide documentation about how to do this reset, sometimes only documenting the procedure to reset a device using a central gateway (coordinator) device from the same brand (which then sends some reset commands to it through the existing network). If you are considering buying a device, it would be useful to look in the documentation in advance for a reset procedure, or contact the manufacturer to ask. Having a reset procedure available avoids making your device unusable by letting it join the wrong network, losing a network key, and so on.

For the Meazon Izy plugs, the reset procedure uses a magnet. Hold it to the side of the device near the LED, which simulates a button press. To confirm that you have the right position, hold the magnet close for a moment and remove it again, which should toggle the switch's state. To reset, hold the magnet close, wait for a few seconds for the LED to blink. It will blink once, then twice, then three times. Then remove the magnet and the device will leave the network. Power-cycle the device to get it to join again.

Note that ZigBee Light Link provides a factory reset procedure by sending special **Touchlink** messages, which works even when the sender and the device to be reset are in separate networks, removing the need for a physical reset procedure. This means that devices that support both ZLL and ZHA (such as the Philips Hue lights) can be used, but it might be hard or impossible to reset them (without also buying a matching ZLL remote control). If XBee allowed sending special inter-PAN messages in the future, there might be a way around this but, at the time of writing, this was not the case.

Discovering services

Now that the device is joined to your network, you can scan it for available services. This serves two purposes:

- Finding out what profile and clusters it actually supports. From the documentation, it is clear that the Meazon Izy plug supports ZHA and it seems safe to assume that it will support commands from the on/off cluster, but using a scan to confirm this helps to prevent problems.

- Finding out the endpoint numbers it uses, so packets are properly addressed. Technically, you can also send messages to all endpoints at the same time by using 0xff as the destination endpoint, but that might have unintended side effects (especially if the device implements multiple profiles or device types).

The discovery process itself is too complex to be completely detailed in this book, but there is a complete sketch provided in the code bundle and called ZdpScan.ino. A copy of this example is also provided as part of the xbee-arduino library, which might be more current by the time you read this.

To use the example, upload it to one of your Arduinos with an XBee module attached (check the serial configuration in the sketch for your hardware). The sketch will scan the network for all devices present and show the results on the serial monitor. For example, this might look like this:

```
Discovering devices
0)  0x0013A20040E2C832  (0x4FC6,  Self)
1)  0x0013A20040D85F9D  (0x0000,  Coordinator)
2)  0x00124B0003D2FC0B  (0xB2D2,  Router)
3)  0x0013A20040DADEE0  (0xC5F7,  End device)
Finished scanning
Press a number to scan that node, or press r to rescan the network
```

This shows all devices found on the network, based on the neighbor table each device keeps. This list might also include devices that have been powered down or are out of range, since these tables keep old values around for a while.

The listing shows both the long and short address and the device type, which should help you figure out which is which. The first couple of bytes of the long addresses indicate the manufacturer of the device. Here you can see all 0x0013A2 devices are XBee modules, making device 2 stand out: that is the Meazon Izy plug.

By entering 2 (in the Arduino IDE serial monitor, you also have to press **Send** or Enter), a service discovery scan will start. For the Meazon Izy device, this outputs:

```
Discovering services on 0xB2D2
Active endpoints response
  About: 0xB2D2
  Endpoints found: 0x0E, 0x09
Simple descriptor response
  About: 0xB2D2
  Endpoint: 0x0E
  Profile ID: 0x0104
  Device ID: 0x0000
  Device Version: 00
  Input clusters: none
  Output clusters: 0x0019
Simple descriptor response
  About: 0xB2D2
  Endpoint: 0x09
```

```
Profile ID: 0x0104
Device ID: 0x0009
Device Version: 00
Input clusters: 0x0000, 0x0003, 0x0004, 0x0005, 0x0006, 0xFCBD
Output clusters: 0x0000
```

This is a lot of data, but you should already be able to figure out most of it. The output shows three different responses, corresponding to requests sent out during discovery. The first response contains a list of the active endpoints. This device has two endpoints, using the identifiers 0x0E and 0x09. Based on this response, the sketch has requested details for both of these endpoints, and the responses to those requests are shown.

The profile ID for both endpoints is 0x0104, which is the ZigBee Home Automation profile. This ID is listed on the front page of the ZHA specification; there does not seem to be an official and public list of assigned profile identifiers.

The first endpoint listed, 0x0E, is a bit of a weird one—it advertises the ZHA profile, but the single cluster listed (0x0019) is not listed as a valid cluster in either ZHA or ZCL. The manufacturer indicated that this endpoint is used for over-the-air updating of the firmware in the device, so it can be ignored for this example.

The other endpoint, 0x09, is the useful one here. It lists a device ID of 0x0009 that, within the ZHA profile, means Generic – Mains power outlet (*ZHA* section 5.7).

Looking at the ZHA specification more closely, you will find that all ZHA devices must support the server side of the **Basic** (0x0000) and **Identify** (0x0003) clusters. Some others may be optionally supported by all devices. (*ZHA* section 7.1). Furthermore, a Mains power outlet device must support the server side of the **On/Off** (0x0006), **Scenes** (0x0005) and **Groups** (0x0004) clusters (*ZHA* section 7.4.10). See *ZHA* section 5.8 or *ZCL* section 2.2.2 for the identifiers belonging to each cluster.

Looking at this device, it implements exactly all of these mandatory clusters and one extra cluster 0xFCBD, which is manufacturer-specific (*ZCL* section 2.5.1.3). Manufacturer-specific means its exact meaning has been defined by the manufacturer and is unlikely to be documented anywhere.

For toggling the device, the On/Off cluster (0x0006) looks promising. How this cluster works is not defined in the ZHA spec but in the ZCL specification. The On/Off cluster allows receiving three commands, identified by a single byte: Off (0x00), On (0x01), and Toggle (0x02) (*ZCL* section 3.8.2.3).

The ZBExplicitTxRequest objects

So far, you have been using the **ZigBee Transmit Request** API frame for sending messages, where you set a destination address and message content, and then just sent the message. These messages use XBee-specific profile, endpoint, and cluster identifiers. To be able to send packets to non-XBee devices, using different identifiers, you can use the (somewhat confusingly named) **Explicit Addressing ZigBee Command Frame** API frame.

In the xbee-arduino, these frames are represented by ZBExplicitTxRequest objects, which have all the fields that ZBTxRequest objects have, but with the following additions:

- A source and destination endpoint identifier, indicating what endpoint to send the message to, and where to expect replies. These fields are set using setSrcEndpoint() and setDstEndpoint().
- A profile identifier, indicating the application profile used for this message. This field is set using setProfileId().
- A cluster identifier, indicating the cluster used for this message. This field is set using setClusterId().

Sending a message

Now you know the API frame type to use and the identifiers to put in, the last piece of the puzzle is to figure out what content to put in the message. All commands defined by ZCL use the **general ZCL frame format** for their content (*ZCL* section 2.3.1). This contains a frame control byte, optional manufacturer code, sequence number, command identifier, and optional command payload. The manufacturer code only applies to manufacturer-specific messages, so is not present here. The on/off commands do not have any payload (parameters) of their own, so this leaves you with a three-byte content, which looks like this:

0	1	2
frame control	sequence number	command

To create this content in your code, start the `switchHeating()` function as follows:

```
void switchHeating(bool state) {
  uint8_t frame_control = 0x01;
  uint8_t seqnum = 0;
  uint8_t cmd = state ? 0x01 /* On */ : 0x00 /* Off */;
  uint8_t content[] = {frame_control, seqnum, cmd};
```

Here you see these three bytes in the `content` array. The frame control byte value is `0x01`, indicating this is a cluster command, sent from client to server (see *ZCL* section 2.3.1.1 for details, but beware that the bits are shown LSB-first, not MSB-first as is more common). The sequence number can be used to match responses to commands but is always set to 0 here. The command must be one of the three values mentioned earlier (*ZCL* section 3.8.2.3), though only the **Off** (`0x00`) and **On** (`0x01`) commands are used in this code.

With the content created, you can now construct the **Explicit Addressing ZigBee Command Frame** API frame and send this packet to the correct endpoint by ending the function with:

```
  XBeeAddress64 addr(0x00124B0003D2FC0B);
  ZBExplicitTxRequest txRequest(addr, content, sizeof(content));
  txRequest.setSrcEndpoint(0x01);
  txRequest.setDstEndpoint(0x09);
  txRequest.setClusterId(0x0006 /* On/Off */);
  txRequest.setProfileId(0x0104 /* HA */);
  xbee.send(txRequest);
}
```

Here, you see most of the values you previously figured out. Note that there are two endpoint values: the source and destination endpoint. The destination endpoint must match the value previously discovered in the scan, while the source endpoint does not really matter, though it cannot be `0x00` (reserved for ZDP), `0xE6`, or `0xE8` (used by the XBee module itself—try scanning one of your XBee modules to confirm). Do not forget to also update the `addr` variable to the 64-bit address of the ZHA device you are using.

Note that this does not yet update `heating_state`. In the previous example, you updated it when the packet was successfully transmitted. This would work here as well, but there is something better: a ZHA device sends a **default response** message to confirm whether any command was successfully received and executed (*ZCL* section 2.4.12.2). Waiting for that response message makes the system more reliable and even capable of handling situations where the command packet was received but could not be executed for whatever reason.

Feel free to add a line updating `heating_state` here to test your sketch as it is now; however, to reliably complete it, you will have to look into receiving messages first.

The ZBExplicitRxResponse objects

To receive messages, so far you have been reading **ZigBee Receive Packet** API frames from the XBee module. These contain the sender address and the data content, but no indication of the endpoints, profile, or cluster being used for the message. This works fine when exchanging data between XBee modules, all just using the default identifiers, but this will not work when talking to other devices, where you need to know the endpoint, profile, and cluster identifiers to know how to interpret the data received.

To support this, XBee modules have a **ZigBee Explicit Rx Indicator** API frame, which includes the extra details needed. In the xbee-arduino library, these frames are represented by **ZBExplicitRxResponse** objects. To process these objects, you can register a callback using `onZBExplicitRxResponse()`.

These objects support all of the methods that `ZBRxResponse` has, but also add a few methods:

- `getSrcEndpoint()` and `getDstEndpoint()`: Returns the endpoint identifiers from the message
- `getProfileId()`: Returns the application profile identifier from the message
- `getClusterId()`: Returns the cluster identifier from the message

Receiving messages

Whether an XBee module sends **ZigBee Receive Packet** or **ZigBee Explicit Rx Indicator** API frames is determined by the `AO` configuration value. By default, it is set to `0` but, for this example, you will need to set `AO=1` to enable the use of **ZigBee Explicit Rx Indicator** frames.

So far, you have been making configuration changes through the XCTU program, but it is also possible to let the Arduino make such changes. By default, these changes are not written to non-volatile storage, meaning that when the XBee module is power-cycled, the changes are forgotten again. By keeping the stored `AO` value at `0` on all your modules and only temporarily setting it to `1` in the sketches that need it, you can still easily exchange your XBee modules and sketches without having to worry about the `AO` being correct for the current sketch.

To set AO=1 in your sketch, add the following bit to the setup() function:

```
uint8_t value = 1;
AtCommandRequest req((uint8_t*)"AO", &value, sizeof(value));
xbee.send(req);
```

Ensure that you to put it before the call to switchHeating() since that relies on AO being set, but after the XBee setup (including the printErrorCb setup, so any errors while setting AO will be printed).

With AO=1 configured, the coordinator is now ready to receive messages from non-XBee devices. Unfortunately, setting AO=1 causes all received packets, including those from other XBee modules, to use **ZigBee Explicit Rx Indicator** API frames. This breaks the existing handling of received packets, which relied on the **ZigBee Receive Packet** API frame and is now no longer used.

The fix is fairly easy: by using the onZBExplicitRxResponse() callback and explicitly checking the endpoint, profile, and cluster identifiers in received packets against the ones used by the XBee modules, you can still receive these messages as before.

This needs two changes. First, in the setup() function, you must make sure to use the onZBExplicitRxResponse() callback register function, instead of the onZBRxResponse() that you used before:

```
xbee.onZBExplicitRxResponse(processRxPacket);
```

Second, the callback function itself must be modified to expect the right type of response object and handle it appropriately:

```
void processRxPacket(ZBExplicitRxResponse& rx, uintptr_t) {
    if (rx.getSrcEndpoint() == 0xe8 && rx.getDstEndpoint() == 0xe8 &&
        rx.getProfileId() == 0xc105 && rx.getClusterId() == 0x0011) {
        // Normal XBee packets
        [ previous code here ]
    }

    DebugSerial.println(F("Unknown or invalid packet"));
    printResponse(rx, DebugSerial);
}
```

Note that the first argument type changed from ZBRxResponse& to ZBRxplicitRxResponse&. The function itself contains the same code as before (omitted in the previous listing), now wrapped in a check that looks at the identifiers in the received message to verify that the message is indeed transmitted by another XBee module, using a regular **Transmit Request** API frame. The values shown have been experimentally determined by looking at the actual packets received, though the endpoints and cluster ID can also be read and changed in XCTU (in the **ZigBee Addressing** section).

 If you do not set AO=1 but messages are being received from non-XBee modules, these will also be received as normal **ZigBee Receive Packet** API frames, so you need to be careful not incorrectly interpret them as a normal message.

Receiving on/off command responses

Now that you have set up proper reception of messages from non-XBee devices, you still need to actually process them. Whenever you send a command to an On/Off cluster, as you did earlier, it will respond with a so-called **default response** command (*ZCL* section 2.4.12). This is not really a command but a response, but all ZCL messages are called commands.

The default response also uses the general ZCL frame format you have seen before. Again, this contains the frame control, sequence number, and command fields, but now also contains a two-byte command payload:

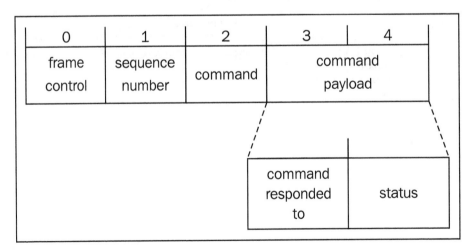

The content of this command payload is specific to the default response command and contains the command that is being responded to, and the status from executing the command (*ZCL* section 2.4.12).

To receive and dissect such a message, a piece of code should be added to the `processRxPacket()` function:

```
if (rx.getProfileId() == 0x0104 && rx.getClusterId() == 0x0006) {
    Buffer b(rx.getData(), rx.getDataLength());
    // Message in the Home Automation On/Off cluster
    uint8_t frame_control = b.remove<uint8_t>();
    uint8_t sequence = b.remove<uint8_t>();
    uint8_t command = b.remove<uint8_t>();
    // frame_control is profile-wide and not manufacturer-specific,
    // command is "Default response"
    if ((frame_control & 0x7) == 0x00 && command == 0x0b) {
        uint8_t command_responded_to = b.remove<uint8_t>();
        uint8_t status = b.remove<uint8_t>();
        if (status == 0) {
            // Switching state worked. Update the heating_state according to
            // the command that was executed (on == 0x01, off == 0x02)
            heating_state = (command_responded_to == 0x01);
            publish(F("House/Heating"), heating_state);
        } else {
            DebugSerial.print(F("Failed to switch plug state. Status: 0x"));
            DebugSerial.println(status, HEX);
        }
        return;
    }
}
```

This code consists of three steps:

1. Verify that this packet uses the ZHA profile (`0x0104`) and the On/Off cluster (`0x0006`). If so, extract the frame control, sequence number, and command from the general ZCL frame header (*ZCL* 2.3.1).

2. Check that the frame is a profile-wide command (as opposed to a cluster-specific command, such as the **On** and **Off** commands you were sending before), is not a manufacturer-specific command, and that it is a default response command (`0x0b`, see *ZCL* section 2.4.12). This code does not also check the sequence number, endpoints, or sender address for simplicity, but you could consider checking those too. Now that the code is sure this is a default response message, the command responded to and status are extracted from it.

3. Finally, the status value is checked. If the switch was successful (a status of zero), the `heating_state` value is updated and the change is published; otherwise an error is printed.

With this, your coordinator should be able to successfully switch the ZigBee power plug remotely and use it to control your heating. The complete sketch is available in the code bundle as `Coordinator_ZHA.ino`.

More ZigBee features

You have now seen how to send basic commands to another type of ZigBee device, using the ZigBee Home Automation profile, and read its responses. Of course, there is a lot more that has not been covered, such as other device types and profiles, reading and setting attributes, subscribing to reports for attribute changes, and so on. These will not be covered in this book, but with the examples and pointers presented here, you should have a firm base for exploring these additional features and you are encouraged to do so.

Summary

In this chapter, you have put control over (a small part of) the world in the hands of your Arduinos. Using your self-built wireless relay, or an off-the-shelf wireless power socket, you have enabled your network to switch things on and off. To keep ultimate control in your own hands, you have added control over the intended temperature to your Beebotte dashboard as well.

These examples should provide a good basis for more complex projects, where the network has more fine-grained control over things than just switching them on or off. Anything you can let your Arduino control (and there are plenty of good tutorials and libraries out there), can be done wirelessly using the same approach used here. Similarly, you could be using physical buttons, knobs, and displays instead of, or in addition to, the Beebotte interface for providing setpoint control, or any other kind of control for that matter.

In the next chapter, you will be looking at using XBee modules without an Arduino or other microcontroller attached. When running them standalone, the XBee modules have some basic, wirelessly accessible I/O options that can remove the need for a full Arduino.

Standalone XBee Operation

5

So far, your XBee modules have been under the direct control of your computer or an Arduino. However, the XBee modules are also capable of running on standalone, which can be useful if you do not need the capability of a programmable microcontroller but just need to remotely read or toggle a few pins.

By using special API frames and radio packets, every XBee module allows reading most of its pins remotely or controlling their output level. Both the digital and analog inputs and digital output is available on all the modules. Some modules also allow you to use **PWM (Pulse-Width-Modulation)** for the output, though typically only on selected pins.

Using these features, you can build a device using just an XBee module, thereby saving the additional cost and size of an Arduino and a corresponding shield. Note that you can also control the pins on an XBee module connected to an Arduino, though most shields leave all these pins unconnected.

In this chapter, you will see how to use the standalone XBee modules for both input and output. The following two setups will be described:

- Using a standalone XBee module to read a switch. This is used to add a window sensor to your system that automatically disables the heating when a window is opened (allowing for the quick and easy ventilation of your house without having to worry about turning the heating down).

- Using a standalone XBee module to control a relay. This is similar to the Arduino-controlled remote relay described in the previous chapter, but removes the need for an Arduino.

For these setups, the recommended hardware is as follows:

- One XBee ZB module (such as `https://www.sparkfun.com/products/11217`).
- One XBee breakout board (such as `https://www.sparkfun.com/products/8276`). Make sure that you also get header pins and XBee sockets in case these are not included with the breakout.
- One SparkFun Breadboard Power Supply USB 5V/3.3V (`https://www.sparkfun.com/products/8376`).
- One breadboard (such as `https://www.sparkfun.com/products/12002`; the mini breadboards are too small).
- Some jumper wires (such as `https://www.sparkfun.com/products/11026`).
- One SparkFun Beefcake Relay Control kit (`https://www.sparkfun.com/products/11042` for the relay setup only).
- One SparkFun Magnetic Door Switch Set (`https://www.sparkfun.com/products/13247` for the window setup only).

This list assumes that you will be building one example at a time, reusing most of the components for the other project. If you want to keep both the examples intact, you should get two copies of the common items.

Here are all these hardware items shown together:

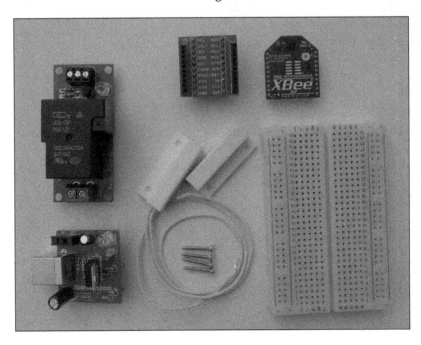

Instead of getting the breakout and power regulator, you could also consider using an XBee Explorer USB, which essentially combines both.

In this chapter, you will see how to wire, configure, and write the code for both of these setups, starting with the window sensor and ending with the relay module.

Creating a window sensor

This example will use a standalone XBee module to read the status of a reed switch in order to detect whether a window is open or closed. In this section, some options for the hardware setup are given. The setup using the recommended hardware items looks as follows:

In addition to setting up the hardware, some configuration is needed to let the standalone module send out the pin state whenever it changes and some modifications must be made to the coordinator sketch to process these messages, all of which will be described next.

Connecting things

To let the XBee module work standalone, you will have to connect a few things to it. As the XBee module pins are fairly short, you cannot just connect the female jumper wires to them, and because XBee uses 2.0 mm spacing, it cannot be directly connected to a breadboard, stripboard, or similar boards (as these use 2.54 mm spacing). You could directly solder some wires to the pins but it is easier to use a breakout board.

A breakout is essentially just a printed circuit board that directly exposes all the pins of a chip (or, in this case, an XBee module) on 2.54 mm-spaced pins, allowing its usage on a breadboard or stripboard. Typically, breakout boards contain no additional components; they just make the connections.

In this example, the XBee breakout from SparkFun is used but there are some near-identical alternatives available.

Powering the module

To run, the XBee module needs a stable 3.3 V power supply, capable of supplying at least 50 mA of current (for the XBee ZB, 802.15.4 and DigiMesh modules, the XBee-PRO modules have higher requirements). In this example, the circuit is built on a breadboard, so it uses a power supply that can be directly plugged in the breadboard, supplying 5 V and 3.3 V from a common USB adapter.

In your final projects, you might want a module like this to be mounted on a bit of stripboard and powered from the mains. As most AC adapters do not provide a well-regulated voltage (which can cause damage to the XBee module when connected directly — refer to `https://www.sparkfun.com/tutorials/103` for some background information), you will need some kind of a regulator to provide a stable 3.3 V.

Alternatively, you could get a breakout with a regulator on board such as the SparkFun XBee Explorer Regulated (which requires a 5 V input that is then regulated down to 3.3 V) to simplify your design. Even simpler to use is an USB-powered board such as the SparkFun XBee Explorer USB, but this also contains the USB-to-serial chip that you will not be using thus making things significantly more expensive.

The options to battery-power the XBee module will be discussed in *Chapter 6, Battery Power and Sleeping*.

Remember that the XBee module runs at 3.3 V and cannot be directly connected to a device that outputs a 5 V signal. No need to worry about this if everything runs at 3.3 V though.

Connecting the window sensor

For a window sensor, a reed switch is commonly used. A reed switch is essentially just a switch that is typically contained in a tiny glass tube. The switch is normally open but closes when a magnet is nearby. By mounting a reed switch on the stationary part of the window frame and putting a magnet on the moving part, you have a switch that will close when the window is closed and open when the window is opened.

Reed switches are available separately, but you can also buy one that has already been prepared to be used on a door or window frame such as the Magnetic Door Switch Set from SparkFun (https://www.sparkfun.com/products/13247).

To allow the XBee module to read the switch state, connect one side of the switch to the DIO1 pin and the other to GND (optionally through a 1 kΩ to 5 kΩ resistor to protect it from a short circuit in case you accidentally configure the pin as an output). This skips the DIO0 pin, which is configured for a commissioning button by default.

You might expect to connect the other end of the switch to the VCC pin; however, using GND lets you take advantage of the the internal pull-up resistors in the XBee module. If you are unfamiliar with the pull-up resistors, you can refer to a good introduction at https://learn.sparkfun.com/tutorials/pull-up-resistors/.

This can be summarized as follows:

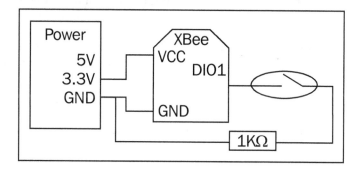

I/O pin naming

On the XBee modules, the I/O pins are typically named ADO/DIO0. This indicates that the same physical pin can be used not only as an ADO pin for the analog input (using an integrated analog-to-digital converter) but also as DIO0 for the digital input and output. In this chapter, the DIO pin names will be used.

Breakout boards typically use either one of these names in their label but of course, you can use the other pin function as well.

The pinout on the different types of XBee modules is very similar but there are some small differences in the capabilities. For a detailed mapping of the pin numbers to the names and functions, see the relevant product manual.

Configuring the XBee module

Now that you have wired up things, it is time to configure the module. Of course, you should first make sure that the module is joined to the network properly, as described in *Chapter 1, A World without Wires*. Once the module is joined to the network, you can insert it into the breakout and the rest of the configuration can be done remotely, as described in the same chapter.

Before you can use an I/O pin on the XBee module, you have to configure the pin mode, just as on the Arduino. By default, most of the pins are set to disabled though some of them have a special function (such as the commissioning button and signal strength indication LED).

The mode of the DIO0 to DIO7 pins can be configured using the D0 to D7 commands and the DIO10 to DIO14 pins using P0 to P4. These commands apply to the XBee ZB modules; the other modules have a different number of available pins and corresponding commands, so be sure to check the product manual.

For this window sensor node, you will have to configure the DIO1 pin as the input, which means setting D1=3. The exact options for these configuration values vary a bit for each pin, but generally they are as follows:

- 0: Disabled
- 1: Pin-specific function (RSSI led, CTS handshaking, and so on)
- 2: Analog input
- 3: Digital input
- 4: Digital output, low
- 5: Digital output, high

To make sure that the internal pullup is enabled on the DIO1 pin, check the PR configuration value. Each bit corresponds to a single I/O pin, so bit 1 needs to be set to enable the DIO1 pullup. The default value of 0x1FFF enables the pullups on all the input pins, which is fine.

Remotely sampling the pins

Now you should have an XBee module with a reed switch connected to an input pin, but it is no fun if you cannot read the state of the switch. Reading the pin states is usually referred to as **sampling** the input pins.

XCTU does not offer an easy single click way to do this, so you will have to dive in a bit deeper. There are basically three ways to (remotely) read the input pins, as follows:

- Query the pin state by sending an IS (input sample) command
- Configure the module to periodically transmit the sample data using the IR (I/O sample rate) configuration value
- Configure the module to transmit the sample data when an input changes using the IC (I/O change detection) configuration value

All three of these can be combined as well. Next, you will first explicitly query the pin state to see if it works. For the final application, you will configure the module to automatically transmit a sample whenever the switch changes its state, which saves the coordinator from having to continuously query the window state.

Querying the pin state

To actively query the pin state, you will have to send an IS command to the XBee module. If the module is directly attached to the computer, you can use the **AT Command** API frame. However, you can also send the command to the window sensor module through another XBee module using the **Remote AT Command** API frame. If you send the latter API frame to a local XBee module, it will forward the command to the remote XBee, which then executes the command and returns a reply.

Building and sending this API frame works similar to the Hello, World! packet that you sent in *Chapter 1, A World without Wires*. So, make sure that you select your locally connected XBee module on the left in XCTU and head over to the **Consoles Working Mode** on the top right. Click on the **Open connection** button to make the console active.

Next up is creating the proper API frame to send. As before, you can click the **Add frame** button to prepare a frame. Use **Frames Generator** to build a **Remote AT Command** frame that looks as follows:

```
                        XBee API Frame generator

 XBee API Frame generator
 This tool will help you to generate any kind of API frame and
 copy its value. Just fill in the required fields.

 Protocol:    ZigBee            ▼   Mode:  API 2 - API Mode With Escapes   ▼
 Frame type:  0x17 - Remote AT Command                                    ▼
 Frame parameters:

   (i) Start delimiter       7E

   (i) Length                00 0F

   (i) Frame type            17

   (i) Frame ID              01

   (i) 64-bit dest. address  00 13 A2 00 40 E2 C8 32

   (i) 16-bit dest. address  FF FE

   (i) Remote cmd. options   02

   (i) AT command            HEX    ASCII
                             IS

   (i) Parameter value       HEX    ASCII

   (i) Checksum              7B

 Generated frame:

 7E 00 0F 17 01 00 7D 33 A2 00 40 E2 C8 32 FF FE 02 49 53 7B


      Copy frame                        Close           OK
```

The **64-bit dest. address** field includes the address of the remote XBee module. You can either put the address of your window sensor XBee here or use the broadcast address, `00 00 00 00 00 00 FF FF`, to query all the XBee modules in the network at the same time. (The modules without any input pins configured will return an error response.)

When you send the API frame that you created by clicking on **Send selected frame**, the target XBee should immediately return a response. This looks something like the following screenshot:

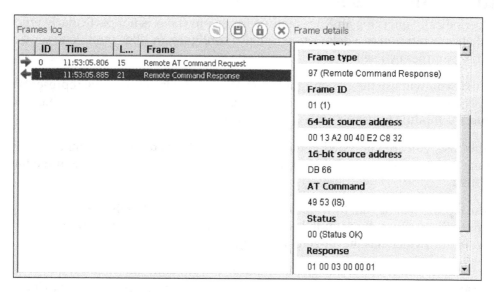

Here, you can see that this is a reply to the `IS` command. The command was executed successfully (**Status OK**) and the return value is `01 00 03 00 00 01`. This return value is the actual sample data, consisting of the following:

- `01`: The first byte is the number of samples in the packet. This is always `1` for the XBee ZB modules, but can be more on the other modules.

- `00 02`: These are the digital pins for which information is present in the packet. Each bit represents one pin, so the rightmost bit is `DIO0`, the next bit `DIO1`, and so on. This value (which is `0000 0000 0000 0010` in binary) indicates that only the `DIO1` pin is present in the packet. This bit is set to `1` for all the pins configured as digital input, regardless of whether the window is open or closed.

- `00`: These are the analog samples present in the packet, which are none in this example.

- • `00 02`: These are the values for the digital pins that have been sampled. Again, the rightmost bit is `DIO0`, so this example indicates that `DIO1` is high (`1`). Try querying the pin state with the window open and closed; you should see this bit toggle.

For more details about this sample format and the format used by other types of the XBee modules, refer to the relevant product manual.

Automatically sending the sample data

As periodically querying an XBee module for its pin states is not particularly efficient (and always introduces a delay), you can also configure an XBee module to send out the I/O samples by itself.

For this, you will use the `IC` configuration value. Each bit in this value represents a pin. The logic is simple: when a bit is set in `IC` and the corresponding pin changes value, the module transmits an I/O sample packet.

In this case, you are interested in the `DIO1` pin, which has the switch connected. By setting `IC=2` on the remote XBee module, any changes to this pin will be immediately reported.

These reports will be sent to whatever address is configured in the `DH` and `DL` configuration values. By default, both are `0` and the samples are sent to the coordinator (which has the all-zero address).

If you did not connect the coordinator module to your computer to test but connected another XBee module, it is easier to set `DH=0` and `DH=FFFF` so as to let the window sensor broadcast its samples to all the nodes in the network. Do not forget to rest both to `0` before proceeding to the next section because here, the samples will need to be sent to your coordinator module.

Try toggling the switch a few times. If you set up things correctly, you should see the packets being received in the **Consoles Working Mode** window for the locally connected XBee module. Note that you saw the **Remote Command Response** API frames when manually querying for the data. Now, you will see **IO Data Sample RX Indicator**, which is similar but not sent in response to any command. If you click the latter, you will see that XCTU conveniently parses the data for you and you can, for example, read `DIO1/AD1` digital value directly. If you look at the raw bytes at the top, you might recognize the same data as before; the packet type is different but the actual sample data is formatted in the same way.

The configuration values

If you followed the preceding instructions, your window sensor XBee should now have the following settings:

- D0=2 to configure DIO1 for the input mode
- PR=1FFF to enable the internal pullups on all the input pins
- IC=2 to enable transmitting a sample whenever the DIO1 pin changes the value
- DH=0 and DL=0 to let the samples be transmitted to the coordinator

This list does not show the settings that are needed to join the module to your network, as listed in *Chapter 1, A World without Wires*.

Receiving the samples on the coordinator

Now that you configured the standalone window sensor XBee module to send all the changes to its input pin to the coordinator, you can modify the coordinator sketch to receive them. You can continue working on top of the coordinator sketch from the previous chapter, which will receive a few bits of new code. The Coordinator_Window.ino example in the code bundle is based on the Coordinator.ino example from *Chapter 4, Controlling the World*.

The ZBRxIoSampleResponse objects

In the xbee-arduino library, the ZBRxIoSampleResponse objects are used to represent the **IO Data Sample RX Indicator** API frames. To process them, you can register a callback using onZBRxIoSampleResponse().

These objects are similar to the ZBRxResponse objects and offer the same methods and also add some methods to access the sample data. The most interesting ones are as follows:

- isAnalogEnabled(): This accepts a pin number and returns whether an analog sample for that pin is included in the response
- isDigitalEnabled(): This accepts a pin number and returns whether a digital sample for that pin is included in the response
- getAnalog(): This accepts a pin number and returns the analog value for that pin
- isDigitalOn(): This accepts a pin number and returns the digital value for that pin (true means high and false means low)

These objects have some additional methods; refer to the xbee-arduino documentation at https://github.com/andrewrapp/xbee-arduino for a full list.

The pin numbers passed to these methods refer to the I/O pin number (so 0 means DIO0) and not the physical pin numbers on the XBee module.

This same object is used for the DigiMesh and 868 modules while XBee 802.15.4 modules use an Rx16IoSampleResponse or Rx64IoSampleResponse instead.

Receiving the I/O samples

Before you can receive the samples, you will need some place to store their values. Add the following global variable for this:

```
bool open_windows[NUM_ROOMS] = {false, false};
```

This adds an open_windows array like the temperatures array, which keeps a window state for each room. Rooms without a window sensor are not a problem; they will just remain false (closed) forever and not influence the heating decision in any way.

To receive these I/O sample messages, you will have to add one more callback to the setup() function:

```
xbee.onZBRxIoSampleResponse(processIoSample);
```

This registers a callback, which must be defined too:

```
void processIoSample(ZBRxIoSampleResponse& rx, uintptr_t) {
  XBeeAddress64 addr = rx.getRemoteAddress64();
  if (addr == 0x0013A20040B1924F) {
      bool open = rx.isDigitalOn(1);
      open_windows[LIVINGROOM] = open;
      inputs_changed = true;
      publish(F("Livingroom/Window"), open);
      return;
  }
  if (addr == 0x0013A20040B19240) {
      bool open = rx.isDigitalOn(1);
      open_windows[STUDY] = open;
      inputs_changed = true;
      publish(F("Study/Window"), open);
      return;
  }
  DebugSerial.println(F("Unknown IO sample"));
  printResponse(rx, DebugSerial);
}
```

This checks the address of the sender in order to find out which window sensor sent a value. It extracts the state of the digital pin 1 (DIO1) from the sample by calling the isDigitalOn() method. This value is stored in the open_windows array as well as published to Beebotte. Do not forget to add a resource in Beebotte for this using the bool type. Similar to when processing temperature values, the inputs_changed variable is set to trigger a re-evaluation of the current heating state. You should, of course, adapt the code to the number of window sensors that you have and use their addresses in the comparisons.

To actually make use of this window status information, add the following snippet to the top of the decideHeatingState() function:

```
for (uint8_t i = 0; i < NUM_ROOMS; i++) {
    if (open_windows[i])
        return false;
}
```

As long as a window is open, the heating will be turned off and remain off until all the windows are closed again. When the last window is closed, the heating will automatically turn on again provided that there is a room below the setpoint, of course.

Beware that this might cause the heating to quickly switch on and off when the window is repeatedly opened and closed, which might not be good for the heating system that you use. You could consider adding some code to prevent it from changing its state when it was already recently switched.

One thing that is not handled by this example is that when the coordinator is reset, it will not know about any of the window states until they change as the window sensors will not transmit their value before that. You could consider fixing this by letting the coordinator send an IS command to every window sensor on the startup and handling the resulting command responses.

When you have enabled AO=1 to receive the data from the non-XBee devices, any sample data received will also use the **Explicit Rx Indicator** API frame and not **IO Data Sample RX Indicator**. This means that you will have to manually check the the endpoint, profile, and cluster identifiers in order to know that the message contains the sample data and you will need to parse the message yourself. The format is the same that you saw in XCTU before. Processing the sample responses in this setup is not covered in this book, but you should have all the tools and know-how to write the required code yourself.

Creating a standalone relay

In addition to reading the input pins remotely, you can also control them remotely. All the XBee modules support a digital output (low/high) and some also support PWM output (pulse width modulation, as used by `analogWrite` in Arduino).

This section shows how to use a digital output pin to control a relay, letting you make a simpler version of the remote relay that you saw in *Chapter 4, Controlling the World*. Obviously, the wiring will be different as there is no Arduino involved anymore. The coordinator sketch will also need some modifications as the XBee module needs a different kind of a message to change its output pin.

When using the recommended hardware items, this example should look as follows:

Connecting things

Most of the wiring is the same as with the window sensor. You will need some kind of a breakout board and power supply. The relay module can be wired up by connecting GND and VCC/5V to the power supply and the relay control pin to the `DIO1` pin on XBee.

Note that most relays need a 5 V power supply as 3.3 V does not provide enough power to toggle the switch. As most of the relay modules contain a transistor to reduce the current load on the control pin, the control pin will work on 3.3 V just fine. As the relay will not output 5 V on the control pin either, the XBee module will not be damaged by this connection.

This can be summarized in the following diagram:

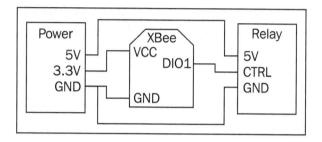

 Just as the pins on the Arduino, the current that the XBee I/O pins can handle is limited. Most pins on an XBee ZB module can only handle up to 4 mA of current—not nearly enough to drive a typical LED at full brightness. All the pins together cannot handle more than 40 mA either. Be sure to consider the current needed by any load that you connect to these pins and check the product manual for your board (under **Electrical characteristics**) to find out how much current each pin can handle exactly.

Configuring the XBee module

The pin mode for DIO1 must be configured by setting D1=4 (digital output: low). Technically, you can leave the pin mode at the default disabled value as setting the pin value later will also use the D1 command (and set the pin in an output mode when it is written to high for the first time). However, configuring the pin mode beforehand makes sure that the pin will be low as soon as the XBee is turned on, forcing the relay off. It also serves to document that the pin is used as an output.

Try toggling the D1 value between 4 (low) and 5 (high) while you are there, you should hear the relay click on and off in response.

Remotely toggling an output pin

Controlling the pin state is done using the same commands as configuring their mode: D0 to D7 and P0 to P4. To remotely send such a command, you can use a **Remote AT Command Request** API frame, specifying the address of the target XBee module, the command to send, and (optionally) a value to set.

For toggling the relay through the DIO1 pin, you will need to send the D1 command with a value of 4 (digital output: low) or 5 (digital output: high). For this, you can use this helper function, setRemotePin(), as follows:

```
// Set remote pin 1 to the given value
bool setRemotePin(XBeeAddress64 addr, uint8_t pin, bool state) {
  uint8_t command[2];
  // Pin <10 use Dx commands, pins 10+ use Px commands
  if (pin < 10) {
    command[0] = 'D';
    command[1] = '0' + pin;
  } else {
    command[0] = 'P';
    command[1] = '0' + (pin - 10);
  }

  uint8_t value = state ? 5 : 4;

  RemoteAtCommandRequest request(addr, command, &value,
                                 sizeof(value));
  uint8_t status = xbee.sendAndWait(request, 5000);
  return (status == 0);
}
```

This figures out what command to use for the pin to be toggled and what value to set. Then, a RemoteAtCommandRequest object is created and transmitted to execute the command. The return value of this function indicates whether the command was successfully executed, as indicated by the status returned by the remote module.

This function, which allows any pin to be toggled, is a bit more general than you will need but this might come in handy with future projects that require more output pins.

With this helper function defined, you can now define a version of the
`switchHeating()` function that uses this helper to toggle the remote relay,
as follows:

```
void switchHeating(bool state) {
  if (setRemotePin(0x0013A20040F51631, 1, state)) {
    heating_state = state;
    publish(F("House/Heating"), state);
  }
}
```

Again, make sure that you replace the destination address with the address of your
relay XBee module.

Summary

In this chapter, you have seen that an XBee module can be used by itself, without an
Arduino connected. You can remotely change its configuration, read the digital and
analog signals on its pins, and control the output voltages.

For simple applications, you can now include a standalone XBee module in your
network thereby saving some costs, allowing for a smaller device, and saving power.

In the next chapter, you will look at making your modules battery-powered and
exploring the techniques to radically reduce their power usage.

6
Battery Power and Sleeping

By now, you have created a real wireless sensor network, consisting of multiple nodes each containing some sensors and/or actuators that are controlled by a central coordinator.

However, all of your nodes have to be connected to the mains power, which limits the projects that you can build. You might not have a power outlet easily available everywhere and moving the nodes is cumbersome.

In this chapter, you will explore the various options for making your nodes battery-powered, allowing for a greater flexibility in deploying your projects and mobile or outdoor deployments.

Of course, battery power is not really useful if the battery is depleted in just hours or days. By applying the power saving techniques presented in this chapter, you will be able to let your nodes run for months or even years on a single battery charge.

In the first example, you will make the window sensor from *Chapter 5, Standalone XBee Operation* battery-powered. In addition to the components that you already have, you will only need two AA batteries and a battery holder (such as `https://www.sparkfun.com/products/9547`).

In the second example, the temperature and humidity sensor from *Chapter 2, Collecting Sensor Data* will be made battery-powered. Some of the components that you previously used will be replaced by more power-efficient alternatives. Here is a full list of the recommended hardware:

- One XBee ZB module (such as `https://www.sparkfun.com/products/11217`).

- One XBee breakout board (such as `https://www.sparkfun.com/products/8276`). Make sure you also get the header pins and XBee sockets if these are not included with the breakout.

- One Arduino Pro Mini board (`https://www.arduino.cc/en/Main/ ArduinoBoardProMini`). Make sure to get the 3.3V/8Mhz version.

- One USB-to-serial converter, ideally with FTDI pinout (such as `https:// www.sparkfun.com/products/9873`). Be sure to get a 3.3V version to match the Pro Mini.

- One DHT22 (or DHT11) temperature and humidity sensor (`https://www. adafruit.com/products/385`).

- One 10kΩ resistor (optional).

- One breadboard (such as `https://www.sparkfun.com/products/12002`, the mini breadboards are too small).

- Some jumper wires (such as `https://www.adafruit.com/products/153`).

- One Lithium Polymer (LiPo) battery (such as `https://www.sparkfun.com/ products/339`).

- One LiPo charger (such as `https://www.sparkfun.com/products/10217`).

Instead of getting the LiPo battery and charger, you could consider using three non-rechargeable AA batteries or four rechargeable AA batteries and an appropriate battery holder.

Here are all the recommended items shown together (common items are shown only once and the LiPo charger is not shown):

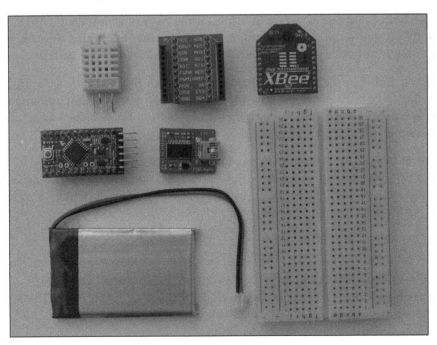

Battery power

Even though it sounds simple, making your Arduino battery-powered is a complicated subject. There are a lot of different types of batteries and multiple ways to connect them, each with their own requirements and effects on the power that is used.

This section gives you a very concise summary of the battery terminology, intended to be exactly enough to make sense of the rest of this chapter. For more detailed information about the batteries and an overview of some common types that are available, see https://learn.adafruit.com/all-about-batteries.

Batteries are devices that store energy and can be used to power electrical devices. Every battery has a nominal voltage (such as 1.5V for alkaline batteries), indicating an average voltage of the battery over its entire lifetime.

Battery capacity is measured in mAh, or milliampere-hour. A battery of 1,000 mAh can supply 1,000 mA for 1 hour, 100 mA for 10 hours, 10 mA for 100 hours, and so on. To estimate the lifetime of a battery, divide its capacity (in mAh) by the average current usage (in mA) to get the lifetime (in hours).

For example, if you use a 1,000 mAh battery and draw 25 mA on an average, it should last about 40 hours. This will give you a useful estimate of the battery's lifetime but it is not a perfect prediction, especially when the current is high (batteries are less efficient for high currents) or the current is low (where the self-discharge rate becomes significant).

All batteries also suffer from **self-discharge**. A battery loses a bit of charge over time, even when no current is drawn from it at all. Self-discharge is nearly negligible for alkaline batteries, which last many years. Among the rechargeable batteries, Li-ion and LiPo batteries have the least self-discharge, though not nearly as low as the alkaline batteries.

Lithium-ion and lithium polymer batteries

These types of batteries are very popular and powerful to use in phones and other equipment and also for Arduino-based projects. They also require some care and so they deserve some special mention here.

Both the names refer to essentially the same type of battery, but a lithium-ion (Li-ion) battery is typically contained in a hard casing, while a lithium polymer (LiPo) battery is contained in a soft (polymer) pouch.

The nominal voltage is usually 3.7 V, with the actual voltage ranging from 4.2 V when fully charged to 3.0 V when fully depleted. Charging these batteries requires a special charger, which is often integrated in devices and also in some Arduino boards.

For more detailed information about these battery types, see `https://learn. adafruit.com/li-ion-and-lipoly-batteries`.

Batteries contain chemicals and store (large) amounts of energy. If not handled properly, they can form a fire hazard or even explode. Always make sure to use a charger that is suitable for the type of battery and observe the manufacturer guidelines and current limits.

Li-ion or LiPo batteries can be especially dangerous when a short circuit, overcharge, or overdischarge occurs, so be sure that an appropriate protection circuit is integrated in the battery to protect against these. Puncturing or rough handling Li-ion or LiPo batteries can also cause internal short circuits and fire, so handle these batteries carefully.

Regulators

Voltage regulators can convert power from a range of input voltages to a fixed and stable output voltage. This can be used to stabilize an unstable input voltage or when a specific voltage is needed that is not otherwise available.

The two most common types of regulators are **linear regulators** and **switching regulators**. Linear regulators can only lower the voltage and operate by converting the excess voltage to heat. Switching regulators can be designed to either lower or raise the voltage and typically operate by very quickly switching between charging a coil and letting the coil charge a capacitor, resulting in the intended voltage.

All the regulators on the Arduino boards are of the linear type, which is also the simplest to include in a design. If you apply power through the barrel jack connector or through the VIN pin (sometimes labeled RAW), you are using the integrated regulator. Usually, you can also bypass the regulator by applying (regulated) 5 V or 3.3 V to the relevant pin (typically labeled 5V or VCC) directly.

Most mains adapters provide an unstable voltage, requiring the use of some kind of regulator for a stable operation and to avoid damaging the components. (Refer to `https://www.sparkfun.com/tutorials/103` for some background on these unstable power supplies.)

Batteries typically have a reasonably stable voltage in the short term, so you can often power a project from a battery directly. A regulator will use a bit of the power for itself, so leaving it out and powering directly from a battery can be a good way to save some power. However, as the battery voltage can vary significantly depending on the battery's state of charge, this only works when all the components to be powered can run on a wide enough range of voltages, which is not always the case (as you will see later).

Saving power

To allow the running of your project on batteries for any meaningful time, you will have to take measures to reduce the current usage. The temperature sensor that you created in *Chapter 2, Collecting Sensor Data*, draws around 80 mA. Suppose you run it on some rechargeable AA batteries rated at 2,000 mAh. This means that the batteries are expected to last about 2,000 mAh / 80 mA, which equals 25 hours.

It is sufficient to say that a wireless sensor network is not so useful when the batteries have to be changed every morning. Fortunately, there are a lot of things that can be done to save power.

The first step in reducing the current usage is to find out where all this current is going. On this sensor, the current usage is roughly divided as follows:

- Main microcontroller: 10-15 mA
- 16U2 microcontroller (serial-to-USB): 15-25 mA (even when not using USB)
- Power LED: 6 mA
- 5 V regulator: 5 mA
- 3.3 V regulator on the Arduino: 0.1 mA (more when loaded)
- 3.3 V regulator on the XBee shield: 0.5 mA
- XBee ZB module: 30 mA
- DHT22 sensor: 1 mA

These values are estimations that are partly based on datasheets and schematics and partly based on (very rough) measurements, but at least the order of magnitude should be correct.

As you can see, the bulk of the current is used by two different microcontrollers and the XBee module; however, something as trivial as a power LED takes up a significant amount of current.

There a few things that you can use in your Arduino sketch to save power. When the full bag of tricks is applied, the main microcontroller's power usage can be reduced to less than 1 µA while it is sleeping (that is, less than 10,000 times of its regular usage). Unfortunately, applying the same tricks to the 16U2 microcontroller is a *lot* more complex, and even if you apply them, the power LED would still negate a large part of the savings anyway.

This means that the Arduino Uno is not particularly suitable for battery operation. Later in this chapter, you will see how the Arduino Pro Mini can be used to get a significantly lower power usage.

Power saving techniques

So what can you do to reduce the power usage of your project?

- **Put the microcontrollers to sleep**: When a microcontroller is not doing anything useful (such as waiting for time to pass or waiting for a pin to become high), putting it to sleep not only saves a lot of the current but also prevents the microcontroller from doing anything. Sleeping will be detailed later in this chapter.

- **Reduce the operating voltage**: When running at a higher voltage, a microcontroller needs more current to keep its transistors running.

- **Reduce the clock frequency**: Every time a transistor switches in a microcontroller, it uses a bit of energy. When running at a higher frequency, transistors will be switching more often, thus leading to more current usage.

- **Prevent floating pins**: When a pin floats, its voltage will vary depending on the static charges and other influences. This can cause the digital input logic to repeatedly switch between detecting a high and low value, causing extra current usage.

  The easiest and safest way to prevent a floating pin is to enable the internal pull-up. (So, if you accidentally connect GND or VCC to the pin, you will not cause a short circuit). Doing this saves power only when actually running as the input pin logic is automatically disabled in most sleep modes.

- **Disable things that are not (currently) in use**: Even when you are not using some parts of the microcontroller such as the analog-to-digital converter, serial ports, or timers, there will be a clock signal running for them, which uses a bit of the power. Disabling these clock signals avoids this. Sometimes, it also helps to keep things disabled normally and turn them on only when you need them. This applies to the external sensors as well, which sometimes offer a sleep mode or you can just cut the power to them.

One important thing to realize is that optimizing for power usage is almost always a matter of trade-offs and always very context-specific. Reducing the power usage usually has a cost in terms of the features (you cannot use a component that is turned off) or time (turning something on when you need it typically takes some time). Sometimes, applying a technique might actually increase the power usage as well. Consider lowering the operating frequency, which lowers the energy needed per second. However, if the lower frequency means that the microcontroller needs to stay awake for a longer time so as to complete an operation, it might end up using more energy after all.

This shows that the best solution is heavily dependent on the details of the project. Giving general advice that works for all the projects is generally impossible. Instead, you will need to have a good understanding of your project and the power saving techniques available and decide on the best approach on a case-by-case basis.

This chapter gives a fairly broad overview about power saving and sleeping. The complete details about the sleep modes and power saving options are available in the microcontroller datasheet. Another excellent piece of documentation written by *Nick Gammon* is available at `http://www.gammon.com.au/power`.

As power saving and sleeping are very specific to the hardware that is being used, this book only considers the Arduino boards based on the Atmel AVR / ATmega family of microcontrollers. A lot of these concepts will apply equally to the other families and architectures but the details and code will differ.

Knowing what to optimize and when to stop

As you can see, there are a lot of power optimizations that are possible. Both the hardware and software changes come at a cost in terms of the engineering effort as well as reduced performance, increased startup times, and so on.

So you might (should!) occasionally wonder: is it worth optimizing this further? Often, you can roughly predict how much a potential optimization would be gained and you should not bother trying to optimize the components that only account for an insignificant part of the total current draw.

For example, if your project is about powering a string of Christmas tree lights that draw 750 mA and your microcontroller draws 3.5 mA, there is not really any point in reducing the latter. (Even if you would succeed in making it zero, you would only have saved 0.5%.)

This also applies to your battery's self-discharge. Once your power usage comes near or below the self-discharge rate, further optimizations might not be worth it.

Also, remember that it is mostly the average current that counts: you can try finding a more power-efficient sensor for your system, but if the sensor is only powered for half a second in every five minutes, it is unlikely to help much compared to reducing the microcontroller's sleep current during the rest of those five minutes.

When it comes to figuring out where it is best to optimize, remember that it all depends on how your project is built. This chapter provides some ways to save power and shows how much current is saved, but the exact numbers will likely be different in your setup, depending on things such as frequency, voltage, temperature, attached sensors, regulators used, and so on.

When trying to estimate the current usage of the various parts of your project, be sure to get familiar with the datasheets of all the components that are involved. These usually provide indications of the current usage in various states. Especially, the AVR microcontroller datasheets provide detailed information (in the **Typical Characteristics** section) that is usually very close to the actual values.

XBee power-saving

Typically, a radio is responsible for a big part of a device's current draw. Even though the XBee radio modules are designed to be low in power, their maximum power usage is typically in the 40-80 mA range while receiving. The XBee modules have the same current usage while transmitting, with the long range XBee PRO modules needing up to 500 mA.

Fortunately, all these modules can be put to sleep just like your microcontroller, bringing their power usage in or below the µA range (less than 1 µA for the XBee ZB modules).

Of course, when an XBee module is sleeping, it will be unable to receive any messages that are addressed to it. This creates a number of problems that need to be solved by the networking stack. In a ZigBee network, this is done by introducing a new class of devices called **end devices** (in addition to the coordinator and routers that you saw before), which are allowed to sleep.

To allow these end devices to sleep, some things work differently for them than what you have seen so far:

- The end devices cannot be routers in the network, which means that they will never forward messages to make them reach their destination.
- When an end device joins a network, the router or coordinator that helped it to join the network becomes the parent of the end device.

- When an end device needs to send a message, it will not try to find out the route to the destination itself but instead just forward the message to its parent, letting the parent figure out the route instead.

- When a message is sent to an end device, it will be routed to its parent. The parent then saves the message and waits for the end device to wake up.

- When an end device wakes up, it sends a **poll request** to its parent to find out if there is any data waiting for it. If so, the end device keeps receiving for a while longer and the parent transmits the pending packet.

- While an end device is awake, it regularly sends a poll request (every 100 ms by default), disabling the radio in between. This lets an end device save power even without actually sleeping.

There are some additional limitations that you should be aware of regarding end devices, which are as follows:

- A parent device has limited buffer space. If the nodes are asleep for long or data is being sent frequently, the buffer space may be filled up and the packets may not reach their destination.

- A parent device has a fixed size child table, limiting the amount of children that it supports. (Routers support 12 children and the coordinator supports 10 in the current firmwares.) When a parent device's child table is full, it will no longer allow any new devices to join the network. Other parent devices can still allow the new device to join, provided that it is in the range of the new device, of course.

- When an end device does not send a poll request for too long, the parent device will remove it from the child table, freeing up space for the new devices.

- When an end device can no longer reach its parent device or if it was removed from the child table (after being powered off for a while), it will automatically try to rejoin the network, possibly through a new parent. This uses a special rejoin procedure, which can work even when the regular joining is disabled using the NJ value. For this to work, it is important for the end device's NJ (node join time) parameter to match the value that is configured in the routers. If the end device has NJ=0xff, it assumes that it can join the network normally. If it has another value, it knows that regular joining is disabled and will use a specialized rejoin procedure, which allows a device to join the network even when joining is normally disabled — if it has ever joined the network before.

  When this happens, the node will get a new 16-bit short address, so keep this in mind if you use these addresses to keep your nodes apart.

This approach applies to the XBee ZB modules and ZigBee networks; other XBee modules solve sleeping in different ways. All the modules support a polling strategy that is similar to the one shown here, though it varies how the **parent** device is defined. In the DigiMesh networks, a synchronized sleep mode is available where the entire network has a synchronized sleep schedule, allowing even the sleeping devices to participate in the network's routing. Refer to the relevant product manual for details on configuring the sleep mode on these devices.

The XBee sleep modes

All XBee modules support two types of sleep, **pin sleep** and **cyclic sleep**, which differ in how the sleeping and waking up is controlled, as shown in the following points:

- When using pin sleep, your Arduino will directly control the XBee's sleep through the SLEEP_RQ pin. When the pin is pulled high, the XBee module will finish any pending operations and then go to sleep, waiting for the pin to go to low again.

- When using cyclic sleep, the XBee module is asleep for most of the time. It will automatically wake up at regular intervals and send a poll request as soon as it wakes up. When its parent has no data pending and no serial data is received, the module quickly goes to sleep again. Otherwise, it waits for a configurable amount of time (which is restarted every time some serial or radio data is received) before going back to sleep. Depending on the required sleep time, you can use a short cyclic sleep (up to 28 seconds) or an extended cyclic sleep (which can cause data loss).

The sleep mode that you can use is configured using the SM configuration value. The valid options are as follows:

- SM=0 means that sleeping is disabled.

- SM=1 means that you can use pin sleep (also called "pin hibernate").

- SM=4 means that you can use cyclic sleep.

- SM=5 means that you can use cyclic sleep with pin wake up. This is identical to cyclic sleep but also lets you wake up the XBee module earlier by driving the SLEEP_RQ pin low.

To let your XBee module operate as an end device, you can either load it with the **ZigBee End Device API** firmware or use the **ZigBee Router API** firmware and set the SM configuration value to a non-zero value. As you will already be running the router firmware on your modules, it is easier to use the latter option. On the XBee ZB S2C modules, only the latter option is available.

To detect whether the XBee module is sleeping or not, the SLEEP and CTS (clear-to-send) pins can be used. When the XBee module wakes up, it makes the SLEEP pin high. A few milliseconds later, when it is ready to receive the serial data, it makes CTS low.

These pins serve two purposes. When the Arduino wakes up the XBee module, the Arduino can detect when the XBee module is ready for the serial data so that it can start sending the data as soon as possible, thus minimizing the power that is wasted while waiting.

When the XBee module is configured for the cyclic sleep mode, the SLEEP pin can be used to wake up the Arduino. By default, the XBee module only makes its SLEEP pin high when it receives data, preventing the Arduino from waking up when there is nothing to do. Alternatively, it can be configured to change the SLEEP pin on every wake up or on every so many wake ups, which can be useful if the Arduino has something to do even when no data is received (such as reading a sensor).

When an XBee module sleeps, it will not respond to the serial data and so it cannot be configured from XCTU. To work around it, you can press the commissioning button to wake up the XBee module and keep it awake for 30 seconds.

When the XBee module is wired up with the flow control pins connected (such as the SparkFun XBee Explorer USB), the XBee will automatically wake up when using pin sleep (SM=1). This works because the SLEEP_RQ pin is also used as the DTR pin and XCTU keeps the DTR pin low while it has the serial port open.

Configuring the network

In addition to configuring the end devices to use the sleeping mode (which will be covered in detail shortly), there are a few things that need to be configured in your coordinator and router nodes. These nodes internally use three timeouts that need to take into account the (maximum) end device sleep time.

- When a node sends the data to another node, the sender waits for the recipient to acknowledge the packet. When no acknowledgement is received after some time, the sender tries resending the packet and eventually gives up and reports failure (this is called the transmission timeout). When sending data to a router node, a small transmission timeout can be used as the recipient is expected to reply immediately.

When sending data to an end device, however, a reply might take significantly longer to arrive depending on the maximum sleep duration of the recipient. As the sleep durations can greatly vary between projects, this transmission duration must be configured correspondingly. (A timeout that is too short would cause the data to be resent or the transmission to fail too soon, but a timeout that is too long would keep a sender waiting needlessly.)

- When a parent receives data for one of its children, it will be stored while waiting for the device to wake up. If the child does not come to collect the pending packet within the **packet buffering timeout**, the parent will remove the packet to free up the precious buffer space.

- When a parent does not hear from a client for longer than the **child poll timeout**, it assumes that the child is dead or has moved out of range and removes it from its child table, freeing up space for new children to join the network.

All these three timeouts can be (indirectly) configured using the SP and SN configuration values. There are the following two cases to consider:

- When the maximum end device sleep time is 28 seconds, SN should be one and SP should be set to the maximum sleep time (in 10 ms units, so 2,800 or 0xaf0 for 28 seconds). All three timeouts will be automatically configured based on this value (see the product manual for the exact calculations used).

- When the maximum sleep time exceeds 28 seconds (using an extended cyclic sleep period or pin sleep), SP and SN should be set such that SN multiplied by SP equals or exceeds the maximum sleep time.

The transmission and packet buffer timeouts are calculated exclusively from the SP value, limiting their value to (a bit more than) 28 seconds. If a node sleeps longer than this, the packets sent to it will sometimes fail to be delivered. It is up to the application developer (that's you!) to let the sleeping node notify others when it is awake in case the other nodes need to send data to it before it returns to the sleep mode.

To calculate the child poll timeout, both SP and SN are used. This allows a child to be asleep for an extended amount of time, and even though data reception can sometimes fail when the node sleeps, it makes sure that the child will remain registered in the parent's child table, thus preventing the child from having to go through the entire (re)join procedure whenever it wakes up.

ZigBee Pro specifies that the **packet buffering timeout** must be at least 7.5 seconds. The XBee ZB routers allow you to set this higher up to 28 seconds, but if you are also using non-XBee routers in your network, be aware that they might drop the packets if an end device sleeps for more than 7.5 seconds. If you have non-XBee end devices, these might not work properly if you configure the packet buffering timeout at less than 7.5 seconds.

As the SP and SN values are also used to configure the sleep times on the end devices, configuring the same value for the SP value across the network generally works when using short cyclic sleep periods, while matching up both SP and SN works when using extended cyclic sleep.

To explore how this sleeping works, you will first see how to make the standalone window sensor sleep and save power using the short cyclic sleep mode. The module will regularly wake up to check the window state and notify the coordinator when the state changes.

The sleeping window sensor

To make the window sensor, which you built using a standalone XBee module, sleep, there are no hardware changes needed (other than connecting a battery, which will be explained next). Letting it sleep does require some changes in the XBee configuration.

Out of the two sleep modes, only cyclic sleep is usable as pin sleep requires an external device such as an Arduino to control the sleep state of the XBee.

In the cyclic sleep mode, the XBee module will be waking up at a regular and fixed interval. This introduces a delay in the detection of the window open/close condition. While the XBee module is sleeping, it cannot monitor its input pins for changes. If an input pin changes during sleep, this will be detected as soon as the XBee wakes up again (assuming that the pin has not changed back already, of course). To be able to respond immediately to a pin change, you will either need a microcontroller to wake up the XBee module or use an edge detector circuit to convert every change (edge) to a short low pulse (which can drive the SLEEP_RQ pin to wake up the XBee).

In this case, checking the window state every once in a while is not a problem. Using a sleep time of 15 seconds seems a good compromise between power saving and the response time.

To enable the cyclic sleep mode and set a 15-second interval, configure the following:

- SM=4 (Sleep mode is cyclic sleep)
- SP=5DC (Sleep period of 1500 x 10 ms)

 Remember that all the numerical values in XCTU are configured using hexadecimal values, including the sleep times!

If you apply these changes, you should see that when you open or close the window, it takes up to 15 seconds for a change to the window state to be received by the coordinator.

In this case, no messages are sent to the window sensor node, so it can go to sleep directly after having transmitted its I/O sample. If you send a reply message to the window sensor (for example, to toggle an output pin), you will see that the window sensor XBee might be back to sleep before the reply message is sent. The message would be saved and delivered on the next wake up, 15 seconds later.

If this is a problem, you can use the SO (sleep options) and ST (sleep timer) configuration values to let the XBee module stay awake for a short while after it sends out its I/O sample message. See the product manual for more details.

Battery power

Until now, you have used a breadboard regulator to provide a stable 3.3 V to the XBee module. However, an XBee ZB module can operate from 2.1 V up to 3.6 V. (Refer to the *Power requirements* section of the product manual. Other XBee families have different voltage requirements.) This is a fairly broad range that lets you use a battery directly without any regulator. A LiPo battery can be up to 4.3 V when fully charged so this would damage the XBee module, but two non-rechargeable AA batteries will start around 3.2 V and will be almost entirely depleted when they reach 2.1 V, making them more suitable to power an XBee ZB module. Rechargeable AA batteries will reach 2.1 V sooner, but can also be used.

Power usage

Using two AA batteries at just under 3.0 V, the power usage by the various parts of the system was measured as follows:

- While the window is closed, some current flows through the pull-up resistor and reed switch. According to the product manual, the pullup is 30 kΩ, that results in 3.0 V / 30 kΩ, which equals 100 µA of current.

- While the XBee ZB module is sleeping, its current is less than 1 µA.

- Every 15 seconds, the XBee module wakes up for about 4 ms and draws around 30 mA. This averages to about 8 µA. This will be slightly more if the window was opened or closed as some data is then transmitted, but this will happen infrequently enough to be ignored.

While the window is closed (which it will be most of the time), the total average current usage will be around 109 µA. Assuming that the batteries have a 2,000 mAh capacity, they can last for over two years (less when using rechargeable batteries due to a significant self-discharge over such a period).

If you use a normally closed reed switch (so that the current flows through the switch only when the window is open) and/or add an extra resistor to limit the current flow further through the switch, the current can be reduced even more, allowing for a longer battery lifetime or smaller batteries (such as an AAA or a 3 V button cell).

Arduino power-saving

Now that you have seen how you can reduce the XBee power usage to almost zero, it is time to look at the other power-hungry device: the Arduino. Most of the techniques listed earlier will be applied here, starting with the replacing of the hardware with other hardware running at a lower voltage and frequency and containing less components. Then, the microcontroller sleep mode will be discussed in detail, letting you reduce the power usage even further.

Normally, the Arduino and XBee module will be asleep and the DHT sensor is powered off. Once every five minutes, the Arduino wakes up to take a sensor reading. This happens as follows:

1. The Arduino briefly wakes up and powers up the DHT sensor (which needs about one second to start up before it can reliably measured).

2. The Arduino sleeps for one more second.

3. The Arduino reads the DHT sensor values and powers the sensor off again.

4. The Arduino pulls the SLEEP_RQ pin on the XBee module to low, causing the XBee module to wake up.

5. The module waits for the XBee module to be fully awake by waiting for its CTS (clear-to-send) pin to go to low.

6. The Arduino constructs a packet for the coordinator and sends it to the XBee module.

7. The Arduino waits for the XBee module to reply with a transmission status.

8. The Arduino tells the XBee module to go to sleep again (by releasing the SLEEP_RQ pin) and goes to sleep itself too.

The XBee configuration

On the XBee side, the configuration is fairly simple. To enable the pin sleep mode (and turn the router into an end device if it is still running the router firmware), set SM=2 to enable pin sleep. No further configuration is needed as the sleep mode will be controlled by the Arduino entirely.

Hardware

You have previously seen that the Arduino Uno is not very power efficient, mainly due to the USB-to-serial converter and power LED that cannot be turned off as well as a fairly inefficient regulator.

Fortunately, some boards are more power efficient, such as the Arduino Pro and Arduino Pro Mini boards. These do not have an USB connection of their own and contain a more efficient regulator. There is still a power LED present, but it has a $10k\Omega$ resistor and so the power usage ends up below 0.3 mA. If needed, the power LED is also easy to remove from the board, and in the newest versions of the Pro Mini, there is even a solder jumper that can be desoldered to disconnect both the regulator and power LED for maximum savings when the regulator is not being used.

For the best results, use the 3.3 V/8 Mhz Arduino Pro Mini. Running at a lower voltage and frequency will provide significant savings when not sleeping. These examples should work on an as-is basis on most of the other AVR-based Arduino boards too, such as the Arduino Uno. Of course, the current usage will likely be higher depending on what else is on the board.

As mentioned earlier, SODAQ provides some interesting Arduino-like boards that are intended to run on a battery and/or solar panel, which can also be interesting to look at. Refer to http://shop.sodaq.com/en/.

The Pro Mini does not have its own USB connector and you will need an external USB-to-serial adapter to program it. SparkFun has a good tutorial about programming the Pro Mini at https://learn.sparkfun.com/tutorials/using-the-arduino-pro-mini-33v.

The latest revision of the Pro Mini has a solder jumper (a small blob of solder that can be removed to break a connection) on the board that lets you disconnect both the regulator and power LED. Powering the board directly through the VCC pin is also possible with this solder jumper left intact but then the regulator might draw a bit of a current through its output pin.

Powering the Arduino

As you have seen earlier, the most efficient way to power an Arduino is to bypass its regulator and provide power directly to its VCC pin. However, batteries always have a voltage that varies over time, so this is only possible when the full range of possible battery voltages matches the required voltage of all the components. In this case, you have these components:

- The Arduino Pro Mini needs 2.5 V-6.0 V when running at 8 Mhz (see the *Speed grades* and *Absolute maximum ratings* sections in the Atmega328p datasheet).

- The XBee ZB module needs 2.1 V-3.6 V (see the *Power requirements* section of the product manual). Note that the other types of the XBee modules have different requirements.

- The DHT22 sensor needs 3.3 V-6 V (see the *Electrical Characteristics* section in the DHT22 datasheet).

This leaves an effective range of 3.3 V-3.6 V. No battery will be able to stay within these limits over a full discharge cycle. Two non-rechargeable AA batteries or 3 V button cells will start out too low already, while three (rechargeable) AA batteries or a LiPo battery will start out too high, and so on.

To fulfill the requirements of this setup, this example is powered from a small LiPo battery connected to the regulator on the Pro Mini through the GND and RAW pins. Using the regulator in the Pro Mini adds about 80 µA of extra sleeping current but ensures that a stable 3.3 V is available.

Note that once the LiPo drops below 3.5 V or so, the regulator output will no longer be a clean 3.3 V due to the **dropout voltage** of this regulator. This means that this setup might not empty a LiPo battery completely, but it should work properly during most of the discharge cycle.

If you do not have a LiPo available, any battery that can supply between 3.5 V and 9 V should work such as three non-rechargeable AA batteries or four rechargeable ones (assuming that the total current draw stays below 75 mA).

Hardware connections

If you are using the recommended 3.3 V Pro Mini, the XBee module can be directly connected (no level conversion is needed). An XBee shield typically also contains its own voltage regulator, LEDs, and perhaps a button, which you will not need here. These are absent on the breakout, avoiding extra power usage as well.

Using a plain breakout module as recommended means that you will need to take care of all the required connections, including the supplying of power to the XBee module and connecting the serial pins, which were taken care of by a shield before. Some additional connections are needed to control the XBee sleep. Here is an overview of all the connections that are needed:

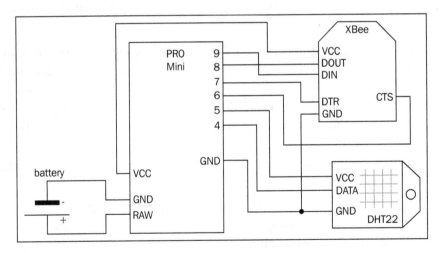

Remember that the crossing lines are connected only if the crossing has a dot. The pins on the Arduino side can be changed to whatever is convenient if you also update the code. Only pins 8 and 9 are fixed as they are dictated by AltSoftSerial.

To be able to turn off the DHT22 sensor, its power input is connected to an Arduino I/O pin. If the pin is set as OUTPUT and written HIGH, it will serve as a 3.3 V power source for the DHT22 sensor. Note that the optional 10 kΩ pull-up resistor is not shown here; it would be connected between the DATA and VCC on the DHT22 sensor.

> In this example, the DHT sensor is powered directly from an Arduino I/O pin. This is possible because the DHT sensor needs only a few mA of current and most AVR-based Arduinos can supply up to 40 mA from their I/O pins. When using a different kind of Arduino or sensor, be sure to double-check the documentation for both so as to avoid overloading your I/O pins! Refer to https://learn.sparkfun.com/tutorials/transistors for more information on controlling a higher current through a resistor.

To be able to control the XBee sleep mode, the DTR/SLEEPRQ and CTS pins are connected to the Arduino. The former lets it control the sleep mode while the latter lets it detect the time when the XBee module has finished its wake up.

With all this wired up, this should look as follows:

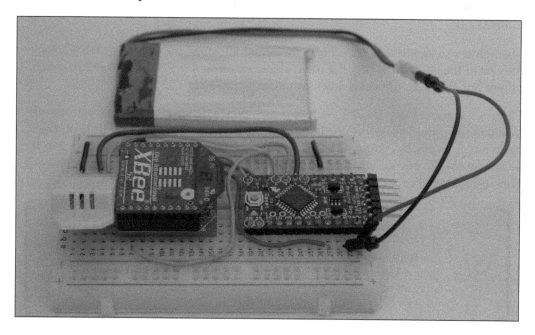

Putting the Arduino to sleep

The official Arduino API functions do not provide any ways to put the microcontroller to sleep. Fortunately, there are some alternatives to allow doing so anyway. The most basic way is to directly access the **hardware registers** (documented in the microcontroller datasheet) or use some **avr-libc** functions to manipulate the registers for you (documented at `http://www.nongnu.org/avr-libc/user-manual/group__avr__sleep.html`). There are Arduino libraries available to make this easier. All of them seem to have some shortcomings but the Adafruit SleepyDog library works well enough for the example in this chapter.

The Adafruit SleepyDog library can be installed through the library manager or downloaded from `https://github.com/adafruit/Adafruit_SleepyDog/`.

Sleep modes

On the AVR microcontroller, sleeping happens by entering one of the many sleep modes.

The deepest sleep mode that is available is called **power down mode**, which stops the main processor clock, effectively powering down most of the chip. In this mode, the power usage for the microcontroller can be reduced to well below 1 μA. Waking up the chip takes some time as the clock has to be restarted. The exact timing depend on the chip, frequency, and settings that are used but it typically takes around a millisecond.

The lightest sleep mode is called **idle mode**, which stops the processor but keeps the rest of the chip, including the main clock, running. This does not save so much power but also does not need any extra time to wake up.

There a few more sleep modes, each of which pick a different point in the trade-off between the power usage on one hand and the features and wake-up time on the other hand. Typically, you will be using only the power down mode but refer to the datasheet for the microcontroller for detailed information about all the sleep modes. The SleepyDog library currently supports only the power down mode.

In most sleep modes, the hardware timer that is used for the `millis()` and `micros()` functions will not be running. When using these functions, remember that their value will not change while sleeping. For the same reason, PWM outputs, which are controlled using `analogWrite()`, will stop working. They get stuck at either a high or low level, depending on the value that they have when the sleep starts.

Waking up

Of course, drawing a little current is nice, but you will need some way to actually wake up from sleep again when something useful needs to happen.

Waking up typically happens after a specific time has passed or when some I/O pin changes its value. In some sleep modes, other **wakeup sources** can be used such as the serial port activity, analog-to-digital conversion completion, or I²C bus activity. All of these are **interrupts**; the microcontroller will wake up whenever an interrupt occurs.

Interrupts are special signals in the microcontroller that signify that some event has happened. For example, interrupts occur when a pin changes from high to low, when a timer reaches a certain value, or when the analog-to-digital converter finishes reading a value.

When such an interrupt occurs when not sleeping, the microcontroller actually stops what it is doing, saves the current state, and starts executing an **interrupt handler** (also called an **interrupt service routine** or ISR). Every type of an interrupt has its own ISR which appears as a special function in your sketch. Once the ISR is finished, the microcontroller restores the previously saved state and continues where it left off.

When an interrupt occurs during a sleep mode, the microcontroller is woken up. The ISR runs and then execution resumes at the point where it was before going to sleep.

Not all the interrupts can be used to wake up from sleep. In most sleep modes, the system clock is stopped to save power (or it is running but only routed to some parts of the chip). Most interrupts will need a running clock to detect the events (such as a timer overflow, serial input or output, or an ADC operation), so these interrupts will not trigger without a clock that is running.

In the power down mode, there are only three kinds of interrupts that can wake up the microcontroller without needing the main clock, as follows:

- The pin interrupts: This includes both the **external interrupts** that are supported by the Arduino `attachInterrupt()` function as well as the **pin change interrupts** (not supported directly though there are libraries available). These interrupts can be used to wake up the microcontroller based on an external event.

- The watchdog timer interrupt: The watchdog is a separate timer that runs on its own internal 128 kHz clock. It allows the waking up of the microcontroller after a configurable interval (typically, 16 ms, 32 ms, 64 ms, 125 ms, 250 ms, 500 ms, one second, two seconds, four seconds, and eight seconds are supported), though its timing is not very precise or accurate.

  This is the wake-up source that is used by the SleepyDog library as well.

- The two-wire address match interrupt: This can be used to wake up the microcontroller when it is used as a slave in the I²C protocol or similar.

Using interrupts will not be covered in detail in this book as the SleepyDog library sufficiently hides them for timed sleeping and only timed sleep is needed for this example. If you need to, for example, wake up on an external pin activity, you will need to use the interrupts for that. Be aware that using the interrupts is prone to subtle and hard-to-detect bugs called **race conditions**, which are often ignored by the various tutorials that are found online. A good article that does talk about them can be found at `http://gammon.com.au/interrupts`. Another good read specifically about sleeping, waking up, and the race conditions involved can be found at `http://www.stderr.nl/Blog/Hardware/Electronics/Arduino/Sleeping.html`.

Creating the sketch

Now that you have hooked up the hardware and know how the Arduino sleeping works, it is time to look at the software to implement all this. The sketch described in this section is called `DhtSend.ino` in the code bundle that accompanies this book. It is quite similar to the example from *Chapter 2, Collecting Sensor Data*, so only the differences between that example and our current example are shown here.

The sketch starts out with some inclusions, of which only SleepyDog is new:

```
#include <Adafruit_SleepyDog.h>
```

Then some global variables and constants are defined:

```
const uint8_t DHT_DATA_PIN = 4;
const uint8_t DHT_POWER_PIN = 5;
const uint8_t XBEE_CTS_PIN = 6;
const uint8_t XBEE_SLEEPRQ_PIN = 7;
```

Make sure that this matches your connections.

In the `setup()` function, the extra pins need to be set up:

```
pinMode(DHT_POWER_PIN, OUTPUT);
pinMode(XBEE_SLEEPRQ_PIN, INPUT);
pinMode(XBEE_CTS_PIN, INPUT);
```

Note that `DHT_DATA_PIN` is configured by the call to `dht.begin()` that is already there.

The call to `sendPacket()` at the end of `setup()` should be removed as `loop()` will start sending a packet right away. The callback registrations are also removed, as explained next.

The interesting things for this sketch happen in the `loop()` function. The general flow of this was described at the start of this section; you might want to refer to that while reading this function. Here is the first half:

```
void loop() {
  // Enable DHT power and give it one second to power up
  digitalWrite(DHT_POWER_PIN, HIGH);
  doSleep(1000);
  // Read data from the DHT sensor.
  dht.read(/* force */ true);
  float temp = dht.readTemperature();
  float humid = dht.readHumidity();
  // Put the DHT back to sleep
  digitalWrite(DHT_POWER_PIN, LOW);
```

This code takes care of powering on the DHT22 sensor, reading its values, and powering it off again. It calls `dht.read()` method, to force the DHT library to get a new value from the sensor. Normally, the library only takes a new reading once every two seconds so as to prevent overloading the sensor. However, while sleeping, the `millis()` counter is not running so the DHT library will think that only a few milliseconds have passed since the previous reading and returns an old reading, even though it is now over five minutes old. Forcing a new reading avoids this problem.

The second half of the `loop()` function handles the XBee-related things, as follows:

```
  // Wake the XBee, and wait until it is awake (up to 1000ms)
  pinMode(XBEE_SLEEPRQ_PIN, OUTPUT);
  if (!waitForPin(XBEE_CTS_PIN, LOW, 1000))
    DebugSerial.println(F("XBee failed to wake up"));
  uint8_t status = sendPacket(temp, humid);
```

```
    if (status == NOT_JOINED_TO_NETWORK) {
      DebugSerial.println(F("Not joined, keeping XBee awake to join"));
      doSleep(30000);
    }
    // Put the XBee back to sleep.
    pinMode(XBEE_SLEEPRQ_PIN, INPUT);
    // Sleep for around 5 minutes
    doSleep(300000);
  }
```

This wakes up the XBee module by setting the DTR/SLEEPRQ pin to low. Before starting to send a packet, this waits until the XBee module is ready for the data as indicated by its CTS pin going to low. This uses a waitForPin() function that will be defined next.

Once the XBee module has fully woken up, the previously measured temperature and humidity are transmitted by the sendPacket() function, the XBee module is put in the sleep mode again, and the Arduino sleeps again for five minutes.

Also note that if sending the packet fails because the XBee module is not joined to the network then some special handling is applied. The Arduino goes back to sleep, but the XBee is kept awake for 30 seconds so that it can try to rejoin the network. This situation can occur when the end device has been out of the range of its parent device for a while, when it has been powered down, or when its parent device is powered down.

XBEE_SLEEPRQ_PIN is not set to HIGH to put the XBee module to sleep. Instead, it is configured as INPUT, allowing the internal pullup on the XBee side to pull up the pin to 3.3 V. This lets the sketch also work with a 5 V Arduino without requiring voltage translators on the DTR/SLEEPRQ pin. (However, DIN and DOUT still require them.)

You might have noticed that this loop() function again uses a blocking style where a single loop() run takes a full five minutes instead of quickly running over and over again. This is acceptable here as the Arduino has nothing else to do in the meanwhile and it makes the code simpler. It also means that xbee.loop() is not (regularly) called, and so the callbacks will not work but these are not really needed in this sketch. If you want to do other things in this sketch as well (such as receiving packets), you might need to rewrite the loop() function using a polling style instead.

In the `loop()` function, a few helper functions were used that still need to be defined. The first is the `waitForPin()` function:

```
// Wait for the given pin to become the given value. Returns true when
// that happened, or false when timeout ms passed
bool waitForPin(uint8_t pin, uint8_t value, uint16_t timeout) {
  unsigned long start = millis();
  while(true) {
    if (digitalRead(pin) == value)
      return true;
    if (millis() - start > timeout)
      return false;
  }
}
```

It is fairly straightforward; it repeatedly checks a given pin and returns as soon as the pin has the required value or a given time has passed (to prevent completely locking up in case the pin somehow never gets the intended value).

Next is the `doSleep()` function that lets you sleep for a specified number of milliseconds:

```
void doSleep(uint32_t time) {
  DebugSerial.flush();
  while (time > 0) {
    int slept;
    if (time < 8000)
      slept = Watchdog.sleep(time);
    else
      slept = Watchdog.sleep(8000);
    if (slept >= time)
      return;
    time -= slept;
  }
}
```

Before going to sleep, the `DebugSerial` stream is flushed. This means that any output that has been queued but not written yet will be written first (as going to sleep would cut short the output).

As mentioned before, the watchdog timer lets you sleep only for a few specific durations (up to eight seconds), so more than one sleep period will be needed to achieve a 5-minute sleep. When calling `SleepyDog.sleep()`, you can pass the number of milliseconds that you want to sleep, but it will always sleep only once, returning the actual number of the milliseconds that it slept.

This code is a bit complex because the SleepyDog.sleep() function only accepts a 16-bit integer as the sleep duration and allows to request only up to 32,767 milliseconds of sleep. As this example needs 300,000 milliseconds (five minutes), the remaining time must be tracked in a 32-bit integer and as this is more than eight seconds, an 8-second sleep is requested.

> The sleep time that is achieved by SleepyDog is not very accurate as the Watchdog oscillator is not very accurate. Additionally, if the Arduino is woken up early during its sleep by an interrupt, the SleepyDog library currently does not detect this and pretends that the full intended sleep duration has passed anyway.

The sendPacket() function is very similar to the function described in *Chapter 2*, *Collecting Sensor Data* but has two important changes, as follows:

```
uint8_t sendPacket(float temp, float humid) {
    // Prepare the Zigbee Transmit Request API packet
    ZBTxRequest txRequest;
    txRequest.setAddress64(0x0000000000000000);
    // Allocate 9 payload bytes: 1 type byte plus two floats of 4 bytes
each
    AllocBuffer<9> packet;
    // Packet type, temperature, humidity
    packet.append<uint8_t>(1);
    packet.append<float>(temp);
    packet.append<float>(humid);
    txRequest.setPayload(packet.head, packet.len());
    // And send it
    uint8_t status = xbee.sendAndWait(txRequest, 5000);
    if (status == 0) {
      DebugSerial.println(F("Successfully sent packet"));
    } else {
      DebugSerial.print(F("Failed to send packet: 0x"));
      DebugSerial.println(status, HEX);
    }
    return status;
}
```

First, the sensor values are passed as arguments instead of being read in this function. This allows the powering down of the sensor before starting the transmission so as to save some power. Second, this waits for and returns the transmit status so that not being joined to the network can be properly handled.

Power usage

With the hardware and software done, you can power up the module and it should start sending the sensor data to the coordinator just as before. In fact, everything should be identical to before except for the power usage, which now looks as follows:

- While sleeping (which is most of the time), the power usage is around 270 µA.

- The DHT sensor is awake for about 1,300 ms every 300 seconds and uses about 1.2 mA. This averages to 5 µA.

- The Arduino is awake for about 600 ms every 300 seconds and uses about 4.5 mA. This averages to 9 µA.

- The XBee module is awake for about 130 ms every 300 seconds and uses about 30 mA. (In reality, the current usage varies a bit due to the polling for the new data.) This averages to 13 µA. The XBee module might be awake for a longer or shorter duration, depending on the network latencies.

In total, this means that the average current usage is 297 µA. Using a 1,000 mAh LiPo battery, this would mean a lifetime of about 4.5 months. (The actual battery lifetime will be shorter than these 4.5 months because of self-discharge.)

What is interesting here is that the biggest part of the average power usage is used while sleeping. This means that sending out the data less often or somehow making the transmission more efficient will not really help much. Instead, replacing the regulator (which accounts for a large part of the sleeping current) will help a lot more.

There are additional power saving tricks that you can add to your sketches that can further reduce the sleeping and running currents. All of these are shown in the DhtSend_Minimal.ino sketch in the code bundle. With this sketch, it should be possible to extend the battery lifetime to over half a year. If you use a more efficient regulator, a significantly longer battery lifetime is possible.

Summary

By now, you will have created a full-blown sensor network consisting of different types of nodes and carrying different sensors and actuators. The network is able to respond to your input and its environment and can control the heating or cooling installation in your house. Some of your nodes are even battery-powered, making them mobile and easy to deploy.

You have all the ingredients to extend on the network that you built or start a completely new wireless project, limited only by your imagination!

Index

Symbols

A

B

O

off-the-shelf ZigBee devices
   controlling 100
   ZigBee protocol 100
on/off thermostats 86
overshoot 89

P

packet buffering timeout 146, 147
packets
   building 58
   constructing, with binary.h file 59
   handling, with binary.h file 58
   parsing, with binary.h file 62
   receiving 62
   sending 58
   sketch, creating 60-64
parent device 144
Personal Area Network (PAN) 25
pin sleep 144
pins, window sensor
   configuration values 127
   pin state, querying 123-126
   remotely sampling 123
   sample data, sending automatically 126
pointers
   about 42
   URL 42
polling method 50
power saving
   about 139, 140
   power optimization 141, 142
   reference link 141
   techniques 140, 141
PowerSwitch Tail device
   URL 87
preprocessor
   URL 42
Processing tool
   about 82
   URL 82
PROGMEM
   URL 42

Pro Mini
   about 151
   URL 151
public profiles, ZigBee protocol
   about 102
   ZigBee Building Automation (ZBA) 102
   ZigBee Health Care (ZHC) 102
   ZigBee Home Automation (ZHA) 102
   ZigBee Light Link (ZLL) 103
   ZigBee Retail Services (ZRS) 102
   ZigBee Smart Energy (ZSE) 103
   ZigBee Telecom Services (ZTS) 102
pull-up resistors
   URL 121
Pulse-Width-Modulation (PWM) 117
Python
   URL 82

R

race conditions
   about 156
   reference link 156
regulators
   about 138, 139
   linear regulators 138
   reference link 138
   switching regulators 138
relay
   about 86
   controlling 97-99
   URL 87
router module 6

S

secure network
   setting up 31, 32
Send on Subscribe (SoS) option 90
sensor data
   collecting 55
   DHT22 sensor, reading 56-58
   hardware setup 36
   packet, building 58
   packet, parsing 62

Thank you for buying
Building Wireless Sensor Networks Using Arduino

About Packt Publishing

Packt, pronounced 'packed', published its first book, *Mastering phpMyAdmin for Effective MySQL Management*, in April 2004, and subsequently continued to specialize in publishing highly focused books on specific technologies and solutions.

Our books and publications share the experiences of your fellow IT professionals in adapting and customizing today's systems, applications, and frameworks. Our solution-based books give you the knowledge and power to customize the software and technologies you're using to get the job done. Packt books are more specific and less general than the IT books you have seen in the past. Our unique business model allows us to bring you more focused information, giving you more of what you need to know, and less of what you don't.

Packt is a modern yet unique publishing company that focuses on producing quality, cutting-edge books for communities of developers, administrators, and newbies alike. For more information, please visit our website at www.packtpub.com.

About Packt Open Source

In 2010, Packt launched two new brands, Packt Open Source and Packt Enterprise, in order to continue its focus on specialization. This book is part of the Packt Open Source brand, home to books published on software built around open source licenses, and offering information to anybody from advanced developers to budding web designers. The Open Source brand also runs Packt's Open Source Royalty Scheme, by which Packt gives a royalty to each open source project about whose software a book is sold.

Writing for Packt

We welcome all inquiries from people who are interested in authoring. Book proposals should be sent to author@packtpub.com. If your book idea is still at an early stage and you would like to discuss it first before writing a formal book proposal, then please contact us; one of our commissioning editors will get in touch with you.

We're not just looking for published authors; if you have strong technical skills but no writing experience, our experienced editors can help you develop a writing career, or simply get some additional reward for your expertise.

Arduino Development Cookbook

ISBN: 978-1-78398-294-3   Paperback: 246 pages

Over 50 hands-on recipes to quickly build and understand Arduino projects, from the simplest to the most extraordinary

1. Get quick, clear guidance on all the principle aspects of integration with the Arduino.

2. Learn the tools and components needed to build engaging electronics with the Arduino.

3. Make the most of your board through practical tips and tricks.

Arduino Robotic Projects

ISBN: 978-1-78398-982-9   Paperback: 240 pages

Build awesome and complex robots with the power of Arduino

1. Develop a series of exciting robots that can sail, go under water, and fly.

2. Simple, easy-to-understand instructions to program Arduino.

3. Effectively control the movements of all types of motors using Arduino.

Please check **www.PacktPub.com** for information on our titles

Arduino Android Blueprints

ISBN: 978-1-78439-038-9   Paperback: 250 pages

Get the best out of Arduino by interfacing it with Android to create engaging interactive projects

1. Learn how to interface with and control Arduino using Android devices.

2. Discover how you can utilize the combined power of Android and Arduino for your own projects.

3. Practical, step-by-step examples to help you unleash the power of Arduino with Android.

Internet of Things with the Arduino Yún

ISBN: 978-1-78328-800-7   Paperback: 112 pages

Projects to help you build a world of smarter things

1. Learn how to interface various sensors and actuators to the Arduino Yún and send this data in the cloud.

2. Explore the possibilities offered by the Internet of Things by using the Arduino Yún to upload measurements to Google Docs, upload pictures to Dropbox, and send live video streams to YouTube.

3. Learn how to use the Arduino Yún as the brain of a robot that can be completely controlled via Wi-Fi.

Please check **www.PacktPub.com** for information on our titles